Physiological Ecology
of Animals

AN EVOLUTIONARY APPROACH

R.M. SIBLY
DPhil
Department of Pure and
Applied Zoology
University of Reading

P. CALOW
PhD, DSc
Professor and Head of
Department of Zoology
University of Sheffield

BLACKWELL SCIENTIFIC PUBLICATIONS

OXFORD LONDON EDINBURGH

BOSTON PALO ALTO MELBOURNE

© 1986 by
Blackwell Scientific Publications
Editorial offices:
Osney Mead, Oxford, OX2 0EL
8 John Street, London, WC1N 2ES
23 Ainslie Place, Edinburgh, EH3 6AJ
52 Beacon Street, Boston
  Massachusetts 02108, USA
667 Lytton Avenue, Palo Alto
  California 94301, USA
107 Barry Street, Carlton
  Victoria 3053, Australia

First published 1986

Set by Setrite Ltd
Hong Kong and
Printed and bound in
Great Britain

DISTRIBUTORS

USA and Canada
  Blackwell Scientific Publications Inc
  P O Box 50009, Palo Alto
  California 94303

Australia
  Blackwell Scientific Publications
  (Australia) Pty Ltd
  107 Barry Street,
  Carlton, Victoria 3053

British Library
Cataloguing in Publication Data

Sibly, R. M.
Physiological ecology of animals: an
evolutionary approach.
1. Physiology   2. Bioclimatology
I. Title   II. Calow, Peter
591.19'1 QP82

ISBN 0-632-01494-6
     0-632-01495-4 Pbk

Contents

Preface

This book is not a classical Physiological Ecology. Rather it is about
the adaptive aspects of resource acquisition and allocation in animals.
There is, of course, a sense in which Physiological Ecology is, by definition, always about adaptation and we shall argue that resource-
allocation patterns are fundamental to most physiological processes. So
what's new? Isn't the subtitle 'An Evolutionary Approach' redundant?
Our hope, in fact, is to signal a shift in emphasis; from an approach
that has been largely comparative, correlational and based on rather
loose *a posteriori* explanations towards a procedure of model building
coupled with an experimental programme of testing the positive explanations and predictions that derive from the models. In this we lean
heavily on optimization techniques borrowed from economics and take
our inspiration for using these in biology from the rapidly and im-
pressively developing fields of Life-history Theory and Behavioural
Ecology.

   To be frank, though, the models are in considerable advance of the
experimental programme — and it is this realization that has partially
motivated the writing of this book. Our aims are to bring the models,
their assumptions and their predictions to working physiologists, ecologists, physiological ecologists and their students in the hope of not
only whetting their appetites for the approach but, probably more im-
portantly, of so challenging them with the lack of evidence as to
stimulate them to design and effect experiments to refute or substan-
tiate what we have proposed. The book can therefore be read at
several levels and in at least three ways. Some readers might find the
first general, introductory chapter and last summary chapter enough!
We have put most of the superficially frightening mathematics into
boxes that some readers might prefer to skip. Other readers, we hope,
might tackle the whole book. We then hope that readers who start
simply might be persuaded to progress ultimately into the boxes! Also,
as aids, we have put all symbols into a glossary at the back of the book
and there are both author and subject indexes. The examples we use

are largely animal, but the issues we raise might apply equally to, and in some cases even be more readily resolved in, plants. So there is still a botanical companion to be written to this work.

We are very grateful to E. Gnaiger, N.K. Jenkins, L. Linton, L. Maltby, H. Møller, K. Simkiss, R.H. Smith, J.F.V. Vincent, C.H. Walker, G.F. Warner and W. Wieser for reading various chapters, and R.M. Sibly wishes especially to thank R.H. Smith for generously discussing many of the topics in considerable depth.

Finally, we hope readers will write to us with comments and criticisms.

R.M. Sibly
P. Calow

March, 1986

1
Introduction

1.1 Aims

There are at least two aspects to every living organism; the one we see, including both structures and processes, and the unseen genetic controls. This is the genotype/phenotype distinction, first made explicit in about 1900 by Johannsen. Study of the form and function of the phenotype can, and indeed does, proceed with only passing reference to genetics. Similarly, the study of the genetic basis of evolutionary change has proceeded with minimum reference to phenotypes.

One area in which the phenotype has to be considered relative to its genetic basis, though, is the study of adaptation. This asks why specific phenotypic traits have evolved in association with specific ecological conditions. This is the kind of question that will be considered below. In particular, we shall be concerned with physiological adaptations of resource acquisition and use because we will argue that this is not only the basis of the way organisms function physiologically but also the basis of their form (allocation of resources between different structures) and behaviour (allocation of resources between different activities). Patterns of resource use are controlled by enzymes and hence ultimately by genes so, according to neo-Darwinian principles, those patterns of use will be favoured that best promote the spread of genes that code for them. This is the sense in which phenotypes are adapted. Nevertheless, one physiological process has to operate within the context of others which constitute the phenotype and this is particularly true given that the resources acquired for use are finite and limited — because the processes and structures involved in this are limited. Hence there are constraints operating within organisms on what patterns of allocation and hence phenotypic traits can evolve. Genotypes determine phenotypes, but the physiology of phenotypes constrains what association of genes (i.e. genotypes) can evolve.

Our aims here are to bring these two aspects of organisms together: to show that understanding the physiology of phenotypes has as much to offer to an understanding of the evolution of adaptations, as natural selection has to an understanding of physiology. The synthesis will be achieved through a class of models commonly used in economics since the problems there are analogous to those confronting all organisms — how best to share limited resources between conflicting demands. This will bring us into contact with optimality theory. In the rest of this chapter, however, we give potted histories of mechanistic models of resource use and neo-Darwinian models of organisms, before introducing the economics models in very general terms.

1.2   Mechanistic models of phenotypes

That organisms build, maintain and reproduce themselves from resources obtained from their food was incorporated into the quasi-biological models of the Greek philosophers, particularly those of the atomists and Aristotle. The atomists even postulated that the bodies of animals represent a dynamic balance between the efflux of atoms from their surfaces and the influx of atoms from their food. The quantification of these processes had, however, to wait until the origin and development of scientific physiology in the eighteenth century. In the 1780s Lavoisier and Laplace declared that the major part of animal heat originated from the oxidative combustion of organic substances in the body and in the 1890s Rubner carried out calorimetric work on dogs and was able to report that the amount of heat produced by these animals over a period of time equals the heat of combustion of the fat and protein catabolized minus the heat of combustion of the urine released over the same period.

These were early days (summarized briefly by Cathcart, 1953) in the application of the **Law of Conservation of Energy** to the energy budget of the functioning phenotype:

$$\text{energy input} = \text{energy output}$$

The energy input derives ultimately from food, but the energy output only equals heat and excreta in steady-state animals. In growing organisms production of tissue has to be taken into account and the more complete equation becomes:

$$\text{food input} - \text{faecal output} = \text{absorbed energy} = \text{production of tissue} + \text{heat loss} + \text{excretory loss}. \tag{1.1}$$

This energy budget model has been used to great effect in the study of animal production, for example by agriculturalists (Brody, 1945) and fish biologists (Winberg, 1956) and in describing the economy of ecological systems at both community (Lindeman, 1942) and population (Phillipson, 1966) levels. Both applied biologists and ecologists have gained considerable insight from energy budget studies on the mechanism of conversion of input to production and the efficiency with which it occurs. The latter is particularly interesting since it gives some information on the extent to which an input to one system can be yielded to another and this is of great importance in agriculture where the recipient is human and in ecosystems where the recipient is the next trophic level.

1.3 Darwinian and neo-Darwinian principles

Darwin's great contribution was not to introduce the concept that life has evolved, nor to prove that evolution is a reality, but to present an objective and refutable hypothesis on the mechanism behind evolution. This account consists of the following logical steps: (**1**) There is considerable, often continuous, variation in traits carried by different members of the same population. (**2**) This variation is derived from a random process (i.e. not related to the direction of evolution) and (**3**) some is heritable (i.e. transmissable from parents to offspring). (**4**) The reproductive potential of individuals within a population is great but (**5**) the resources (space and nutrients) available to the population are finite. Hence (**6**) there will be competition between variants for limited resources and (**7**) those which are most successful (the **fittest**) in this struggle for existence will become most numerous; i.e. will, as Darwin put it, 'survive'. Hence, we have a **natural selection of the fittest**.

A crucial premise in this argument is that of inheritance (3) since though it is conceivable that the outcome of competition might be determined by non-inheritable traits, this would have no lasting effect on the character of the population; in this case the successful traits would die with the individuals that carry them. What natural selection does is to allow successful traits to become more common from generation to generation.

Darwin's own views on the mechanism of inheritance were provisional and, with hindsight, incorrect. Like most of his contemporaries, he favoured blending inheritance, where traits from parents blend in their offspring. But, as early critics of the *Origin of Species* pointed out, this could work against natural selection, diluting more advantageous with less advantageous characters (cf. King, 1984). In fact, Darwin's own realization of this difficulty drove him to embrace a theory which allowed acquired characters to be inherited since this Lamarckian mechanism would enable large advantageous changes to be assimilated quickly before the diluting effect of blending became manifest (Ruse, 1979).

Though it took some time to be appreciated, Mendel's mechanism of heredity was far more compatible with Darwinism than Darwin's own. Mendel postulated the existence of non-blending, particulate factors, now called genes, which determine phenotypic traits and which are transmitted from parent to offspring. Genes may not be expressed in a particular generation and may even be lost by natural selection but they never blend with other genes.

Mendel's discoveries on how factors were transmitted from parents to offspring also had a profound effect on what was viewed as the unit of selection and on how to assess evolutionary success (fitness). Alleles (replicas of a gene coding for a trait, of which there are usually two per parent) are (a) assorted randomly between gametes (Mendel's First Law), (b) segregated independently relative to the alleles of other genes (Mendel's Second Law which is now known not to be universally applicable — some genes are usually transmitted together, i.e. are linked) and (c) mixed with other genes as a result of fertilization in sexual reproduction. Mendel's two laws are now explicable in terms of the chromosomal basis of heredity and the mechanics of meiosis, as, indeed, is linkage. What they imply is that, in principle, the gene combinations collected together in individuals are dissociable. Genes (and to some extent linkage groups) are the enduring units of evolution and hence the units upon which selection works. Because of a lack of understanding of inheritance, Darwin's theory emphasized the individual, and considered fitness in terms of the survival and fecundity of individuals. On the other hand, the Mendelian reinterpretation (known as neo-Darwinism) emphasized genes and measured fitness in terms of the increase or decrease of particular genes (and linkage groups) in populations; i.e. in terms of gene frequencies. As will be clear from the introductory section, our approach is to emphasize the importance of considering genes, or the traits they determine, operating within the context of organismic constraints and trade-offs.

Initially, the neo-Darwinian extension clarified premise 3 in the Darwinian argument (p. 3). Further clarification came when the histological and molecular basis of heredity was worked out. This also shed some light on premise 2 concerning the all-important source of variation. Non-directed modification of the material of heredity (chromosomes, nucleic acids), i.e. mutation, leads to alteration in the expression of genes and hence in the phenotypic characters they specify.

At the same time, there has been a loosening of the concept of the gene as a physically discrete portion of DNA coding for a specific phenotypic trait. Portions of DNA code for proteins, each of which may be involved in diverse phenotypic processes and structures. Moreover, the genome is far more dynamic than was originally believed and even outside meiosis, genes and gene fragments are capable of moving from one part of it to another. Nevertheless, Mendelian principles continue to facilitate good predictions in breeding experiments and so must be a fair approximation of reality, and the fundamental components of the Darwinian argument (1–7 above) remain intact.

Given the premises in this argument (1–5) the conclusions (6 & 7) follow automatically — otherwise the argument would not be a logical one. What makes it of scientific rather than just philosophical interest, however, is that each of the premises can be subjected to scientific scrutiny and might be proved false. Moreover, though the general outcome — the survival of the fittest — is prescribed by the logic, the particular manifestation will vary from taxon to taxon, niche to niche and habitat to habitat. Hence, the occurrence of particular traits in populations invites analysis and explanation within the neo-Darwinian framework. This is the study of adaptation and is the concern of the so-called **adaptationist programme**. (For an introduction to the expanding literature on the development and details of Darwinism and neo-Darwinism see Calow (1983a).)

1.4 Adaptation

At this point it is worth drawing a distinction between the aims and approaches of two major branches of evolutionary biology: **population genetics**, which has grown up in close association with neo-Darwinism, and this **adaptationist programme**. Given certain simplifying assumptions about populations and genes, and assumptions about the fitness of particular genes, population geneticists make deductions about changes and constancies in gene frequencies. On the other hand, given certain (usually simplifying) assumptions about the genetic basis of a trait, the adaptationist programme is concerned with why it has a particular effect on gene frequencies. Population genetics, then, is concerned with the *effects* of natural selection, and the adaptationist programme with the *causal* basis of these effects.

The core **neo-Darwinian hypothesis** usually adopted by adaptationists is that successful phenotypic traits will be those that **maximize fitness**. In principle this means the ones that maximize survival (S) and fecundity (n) and minimize generation time (t) of organisms that carry the traits. However, for reasons already stated, it will not usually be possible in practice to maximize these **components of fitness** simultaneously. There will be trade-offs and constraints so that **optimization** is often a more important element of the core hypothesis than maximization *per se*.

Even in this context there is no absolute recipe for success. Maximizing fitness will mean different things for different organisms in different ecological cirumstances, and so there is always an important ecological dimension to the adaptationist programme. With this in mind there are at least two ways in which the core hypothesis can be applied. On the one hand it can be used to consider why certain traits have evolved in certain ecological circumstances and, on the other, to consider what traits might be expected to evolve given certain kinds of ecological challenge. These are known as the *a posteriori* and *a priori* methods respectively (Calow & Townsend, 1981a). Though related, these two methods are sufficiently distinct in their approach and shortcomings to be treated separately.

1.4.1 A POSTERIORI APPROACH

After observing that an organism with a certain ecology also possesses traits of a particular kind, some attempt is made, in this approach, to explain the traits *vis-à-vis* the ecology, using the core hypothesis. The assumption often implicit to this exercise is that any and all traits are adaptive and invite explanation under the umbrella of the core hypothesis. Yet, as will be made more clear below, this is not always the case since some traits might arise as a result of organizational constraints (associated with particular taxonomic forms) and others from historical accident.

Lewontin (1978) points to the evolution of the human chin as an instructive example in this context. This is relatively large in adults but absent in infants and in apes. Attempts to explain it as an adaptation were never very successful and now it is considered more as an epiphenomenon of other adaptive processes. There are two growth fields in the lower jaw: the bony dentary field and the alveolar field in which teeth are set. They have both reduced in size in human evolution, but the alveolar field has shrunk faster than the dentary one and this has left the latter jutting out as the chin. It is now believed that this differential shrinkage has adaptive significance for the form of the skull but that the chin is a secondary by-product of it. Another example, this time of the involvement of historical accident in the evolution of morphology, is possibly furnished by the horns of the rhinoceros. The Indian species of rhinoceros has one horn and the African species two. In both cases, these are used for fighting or fending off rivals or predators. It seems unlikely, however, that one horn is specifically adaptive for Indian and two for African conditions. Beginning with slightly different developmental and genetic systems the two species responded to the same selection force in different ways. The desire to see adaptation everywhere is often called the Panglossian Fallacy after Pangloss, a character in Voltaire's 'Candide' who was able to think up the most preposterous reasons for anything that came to his attention.

A posteriori, adaptationist explanations are considerably strengthened if traits are observed to change in different populations of the same or related species occupying obviously and measurably different ecological conditions. This suggests that they are responsive to ecological shifts and are not simply fixed taxonomic or organizational features. Moreover, the changes themselves may suggest the nature of their adaptive significance. The assumption is, of course, that the observed divergences in traits have been effected by natural selection according to the criteria specified in the core hypothesis. Hence, correlational and comparative techniques are important features of a posteriori programmes.

Physiological ecologists working on marine littoral faunas have often indulged in this kind of exercise. This is because they have at their disposal (a), large numbers of taxonomically related species, which (b), occupy ecologically distinct and distinguishable zones in close proximity. Factors a and b are important requirements of this kind of approach. Table 1.1 summarizes some typical data on the form and physiology of limpets occupying different levels on the shore. The greater resistance to and tolerance of desiccation and the smaller foot surface to shell height ratio of Patella vulgata as compared with P. aspera can be ascribed to the fact that the former lives higher up the shore than the latter and suffers a greater frequency of longer exposure to air between the tides.

There are two major limitations with this kind of correlational approach. First, the correlation might still have nothing to do with an adaptive cause. For example, the differences between the limpets in the different zones might be phenotypic but not genotypic (due, perhaps, to the differential effect of wave action and/or food supply on growth). This could be tested for experimentally by artificially altering the environmental conditions (e.g. by transporting groups of individuals to different levels on the shore or growing them under constant,

Table 1.1. Some aspects of the physiological ecology of marine limpets. After Calow 1983a, compiled from Davies, 1966, 1969.

	Patella vulgata	*Patella aspera*
Position on shore	upper levels	low levels
Survival at temperatures above 30°C	good	poor
Rate of water loss from tissues	low	high
Ability to withstand water loss	good	poor
Shell shape	▲	▲

controlled conditions in the laboratory), but this is rarely done. The second, major limitation is that there may be an adaptive cause, but not the one specified in the original explanation. For example, the differences in the shell shape of the limpets might be genotypic, but due to an adaptive response to a predator, itself limited by physical conditions to one part of the shore. Here, the trait differences will be correlated with, but not caused by, differences in exposure to physical conditions on the shore.

1.4.2 A PRIORI APPROACH

The problems associated with the *a posteriori* technique can all be attributed to one philosophical principle; namely that a correlation between variables does not prove a particular cause—effect relationship between them. One correlation might, in principle, be the result of one of many causal factors. In fact this principle bedevils all science, for causal inferences are invariably based upon the association of observations (i.e. correlations) and this means that there can never be complete certainty about cause—effect relationships. All that can ever be hoped for, therefore, is an increase in the confidence that can be placed in inferences about causation, and this is achieved in science by critical tests carried out under conditions which are as controlled as they can possibly be made, i.e. by experiment. Hence, the distinction between (a) formulating explanations *after* observing correlations and not going any further, and (b) first thinking out what possible causal factors might be involved (making hypotheses) and then setting out expressly to test these ideas is, though only a matter of degree and timing, an extremely important one. The latter is what is referred to here as the *a priori* approach.

Hypotheses consist of one to many premises which in association make statements (predictions) about the real world (e.g. the neo-Darwinian hypothesis above). The association of premises to generate predictions is an exercise in logic which can either be carried out in words or in symbols using the formalized logical procedures of mathematics. With simple hypotheses, the predictions are often obvious and straightforward whereas with complex ones, involving many elaborately associated premises, the predictions may be far from obvious and might only emerge after considerable and complex mathematical manipulation. These complex hypotheses are often referred to as **mathematical models**.

What patterns of resource allocation are expected to evolve in particular ecological circumstances and what models will be appropriate? To make predictions here it is necessary to specify (from observation or by hypothesis) the demographic effects of particular patterns of allocation, i.e. in terms of S, n and t (definitions on p. 6). Consider a gene or gene-complex (A) coding for a trait. Whether or not A becomes established in the population will depend upon its ability to spread as compared with other genes. This ability to spread (F_A) is defined by its per capita rate of increase,

$$F_A = \frac{dN_A}{dt} \frac{1}{N_A}, \qquad (1.2)$$

where N_A is the density of A in the population. Hence, gene-determined, physiological traits that bring together combinations of S, n and

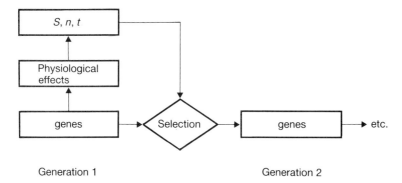

Generation 1   Generation 2

Fig. 1.1. The evolutionary process as an optimizing system in which genes with higher fitness are selected. Not depicted are the effects of mutation, and other random effects. It is important to remember that the effects of genes on S, n, and t may change with changes in environment or in other members of the population.

t that maximize *F* will become more common. This is the **core hypothesis** (above) and is depicted schematically in Fig. 1.1.

It is possible to define *F* precisely in terms of *S*, *n* and *t* as follows:—

$$1 = \tfrac{1}{2}\sum_{t=1}^{\infty} e^{-Ft}S_t n_t \tag{1.3}$$

This is the so-called Euler/Lotka equation and a simple derivation of it is given in Box 1.1. *F* will increase if *S* or *n* increase or if reproduction occurs earlier, but (a) notice that the form of the equation immediately suggests that the timing of reproduction (because it is an exponent) may have more impact on *F* than adjustments in *S* and *n* (see Chapter 6); and(b) remember that there are very likely to be interactions between *S*, *n* and the timing of reproduction, such that these parameters cannot be maximized or minimized indefinitely. We shall return to this later.

Rarely is it possible to carry out exhaustive analyses of physiological processes in terms of demographic effects. Physiological processes operate on a minute-by-minute basis and yet their effects on survival and reproduction are usually more long-term. Physiological causes and their demographic effects therefore operate on different time scales and are often separated in time. It becomes necessary, therefore, to identify more immediate, if less general measures of fitness. To replace, in other words, the demographic parameters with some **phenotypic measure of fitness**. The relationship between the latter and the former is embodied in auxiliary hypotheses which, in a sense, short-circuit between the physiological and genetic level in Fig. 1.1. A typical auxiliary hypothesis that will be used below is the so-called Maximization Principle of biomass production — that natural selection will favour this because by so doing it maximizes growth, and hence minimizes generation time, and maximizes *n*. This is, however, subject to qualification, for example that no extra mortality risks are incurred (see Chapters 2–5). Other auxiliary hypotheses will also be employed in the chapters that follow.

1.5   Economic phenotype

We have considered the mechanism of phenotypic function and the mechanisms of evolution. Those gene-determined, resource-utilization patterns will be favoured that maximize fitness. But given the limited nature of the resources that are made available to metabolism by the

BOX 1.1 Derivation of the theorem that fitness F (i.e. per capita rate of increase) of a dominant gene is given by

$$1 = \tfrac{1}{2} \sum_{t=1}^{\infty} e^{-Ft} S_t n_t \qquad (1.3)$$

where S_t is the probability of individuals carrying the gene surviving to age t, at which age each gives birth to n_t offspring, half of which (on áverage) receive a copy of any allele present in a parent (Table 1.2). As a simplification we·assume both sexes have the same life cycle and individuals mate with other individuals of the same age. We assume that the genetic system is diploid, and that the age structure is stable. Note that equation 1.2 implies exponential increase in gene numbers if F is constant, i.e. there are now e^F times as many copies of a gene as there were one time unit ago.

Suppose at time τ, N_τ copies of a gene are made. Some of these are made from copies in parents of age 1. How many copies of the gene are there in parents of age 1? Well, each parent of age 1 was born at time $\tau - 1$ when in total $N_\tau e^{-F}$ copies of the gene were made. However only a fraction S_1 survived to age 1. Therefore the number of copies of the gene in parents of age 1 is $N_\tau e^{-F} S_1$. And from each of these $\tfrac{1}{2} n_1$ copies of the gene are transferred into offspring (Table 1.2). Hence $\tfrac{1}{2} N_\tau e^{-F} S_1 n_1$ copies of the gene are made at time τ from copies in parents of age 1.

Similarly some copies of the gene are made at time τ from copies in parents of age 2. How many copies of the gene are there in parents of age 2? Each parent of age 2 was born at time $\tau - 2$ when in total $N_\tau e^{-2F}$ copies of the gene were made. However only a fraction S_2 survived to age 2. Therefore the number of copies of the gene in parents age 2 is $N_\tau e^{-2F} S_2$. And each of these transmits $\tfrac{1}{2} n_2$ copies of the gene into offspring. Hence $\tfrac{1}{2} N_\tau e^{-2F} S_2 n_2$ copies of the gene are made at time τ from copies in parents age 2.

Similarly $\tfrac{1}{2} N_\tau e^{-3F} S_3 n_3$ copies of the gene are made

Table 1.2. Demonstration that ratio of copies in offspring to copies in parents is $\frac{1}{2}n$ irrespective of parental genotypes.

Parental genotypes	No. of copies of A in parents	Offspring genotypes	No. of offspring of each genotype	No. of copies of A	Total no. of copies of A in offspring	Ratio of Column 6 (copies in offspring) to column 2 (copies in parents)
Aa × aa	1	Aa aa	$\frac{1}{2}n$ $\frac{1}{2}n$	$\frac{1}{2}n$ 0	$\left.\right\}\,\frac{1}{2}n$	$\frac{1}{2}n$
AA × aa	2	Aa	n	n	n	$\frac{1}{2}n$
Aa × Aa	2	aa Aa AA	$\frac{1}{4}n$ $\frac{1}{2}n$ $\frac{1}{4}n$	0 $\frac{1}{2}n$ $\frac{1}{2}n$	$\left.\right\}\,n$	$\frac{1}{2}n$
Aa × AA	3	Aa AA	$\frac{1}{2}n$ $\frac{1}{2}n$	$\frac{1}{2}n$ n	$\left.\right\}\,\frac{3}{2}n$	$\frac{1}{2}n$
AA × AA	4	AA	n	2n	2n	$\frac{1}{2}n$

at time τ from parents of age 3, and $\frac{1}{2}N_\tau e^{-4F}S_4 n_4$ copies from parents age 4, and so on. Adding these all up we find that the total number of copies of the gene made at time τ, N_τ, is

$$N_\tau = \tfrac{1}{2}N_\tau e^{-F}S_1 n_1 + \tfrac{1}{2}N_\tau e^{-2F}S_2 n_2 + \tfrac{1}{2}N_\tau e^{-3F}S_3 n_3 + \tfrac{1}{2}N_\tau e^{-4F}S_4 n_4 + \ldots$$

Dividing every term by N_τ we get

$$1 = \tfrac{1}{2}e^{-F}S_1 n_1 + \tfrac{1}{2}e^{-2F}S_2 n_2 + \tfrac{1}{2}e^{-3F}S_3 n_3 + \tfrac{1}{2}e^{-4F}S_4 n_4 + \ldots$$

Or, in mathematical shorthand,

$$1 = \tfrac{1}{2}\sum_{t=1}^{\infty} e^{-Ft}S_t n_t \qquad (1.3)$$

Or, in the notation of integral calculus,

$$1 = \tfrac{1}{2}\int_0^{\infty} e^{-Ft} S_t n_t dt \qquad (1.4)$$

This says the same thing as equation (1.3) only more precisely. Instead of taking the intervals of time between possible breeding seasons as 1, the time intervals are made infinitesimally small (equal to dt).

feeding processes and structures and the inevitable conflict between investments — the so-called **Principle of Allocation** alluded to by Cody (1966), Williams (1966) and Gadgil & Bossert (1970) — how do we decide which patterns should be optimum relative to F (or some auxiliary function)? Like so many questions in evolutionary biology, this was probably first posed explicitly by R. A. Fisher in his book 'The Genetical Theory of Natural Selection' (1930) when he wrote:

'It would be instructive to know not only by what physiological mechanism a just apportionment is made between the nutriment devoted to the gonads and that devoted to the rest of the parental organism, but also what circumstances in the life-history and environment would render profitable the diversion of a greater or lesser share of the available resources towards reproduction.'

This is equivalent to the economics problems of allocating limited capital to maximize returns.

Suppose a fraction u_1 of the resources available to an organism is allocated to growth, the remainder going to reproduction (Fig. 1.2). If u_1 is fixed the optimal solution is obtained by choosing u_1 to maximize fitness, F, which may be measured exactly or may be an auxiliary measure. Too high a value of u_1 means that not enough offspring are produced to maximize fitness, but too low a value of u_1 *might* mean that the adult sacrificed growth unnecessarily. In this case (Fig. 1.3) an intermediate value u_1^* maximizes fitness. u_1^* can be found by calculus (Box 1.2).

Now consider what happens if resources are not split simply between growth and reproduction but may also be allocated to defence (Fig. 1.4). Let u_1 and u_2 be the fractions of resources allocated to growth and defence respectively, the remainder going to reproduction. Fitness is now a function of both u_1 and u_2, and is represented graphically in three dimensions (Fig. 1.5a). Following Wright (1932) we shall

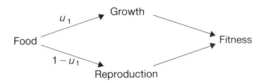

Fig. 1.2. Allocation of resources between growth and reproduction (the simplest model). Both growth and reproduction affect fitness, but each way of allocating resources (described by u_1) results in one fitness value — as in Fig. 1.3.

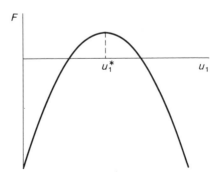

Fig. 1.3. Hypothetical relation between the way resources are allocated (described by u_1 as in Fig. 1.2) and fitness.

BOX 1.2. Finding u_1^*, the value of u_1 which maximizes fitness F.

In Fig. 1.3 F is a mathematical function of u_1, written $F(u_1)$. The slope of $F(u_1)$, $\dfrac{\mathrm{d}F}{\mathrm{d}u_1}$, is horizontal at $u_1 = u_1^*$, i.e.

$$\left.\frac{\mathrm{d}F}{\mathrm{d}u_1}\right|_{u_1=u_1^*} = 0 \qquad (1.5)$$

If the form of $F(u_1)$ is known, equation (1.5) can be used to find u_1^*, e.g. if $F(u_1) = -5u_1^2 + 5u_1 - 1$ then

$$\frac{\mathrm{d}F}{\mathrm{d}u_1} = -10u_1 + 5 \text{ and } \frac{\mathrm{d}F}{\mathrm{d}u_1} = 0 <=> u_1 = \tfrac{1}{2}, \text{ suggest-}$$

ing that $u_1^* = \tfrac{1}{2}$. Note that it is necessary to check that $u_1 = \tfrac{1}{2}$ is a maximum, not a minimum of F.

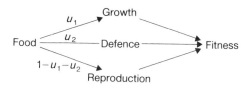

Fig. 1.4. Allocation of resources between growth, defence and reproduction (cf. Fig. 1.2).

call this a **selective landscape**. Fitness is the analogue of 'height above sea level' in an ordinary landscape, and the optimal strategy corresponds to the highest point on the landscape. If there are several peaks in the landscape then individual peaks are called **local optima**, and the highest peak is said to be the **global optimum**. The values of u_1 and u_2 that maximize fitness are designated u_1^* and u_2^* and constitute the optimal strategy.

Just as an ordinary landscape can be represented as a contour map so can a selective landscape, the contours being lines joining points having the same fitness — what we call a **fitness contour**. The contour map corresponding to Fig. 1.5a is shown in Fig. 1.5b.

Now suppose that the allocation of resources to reproduction is fixed — perhaps at the optimal level — so that $u_1 + u_2$ is fixed. Thus for some constant, b

CHAPTER 1

(a)

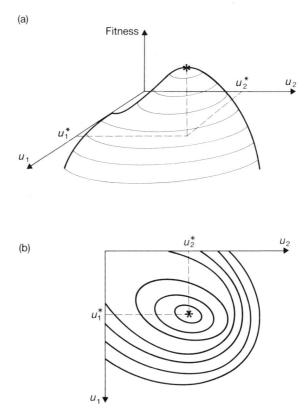

Fig. 1.5. (a) The selective landscape, in which fitness is plotted as a function of the way resources are allocated, described by u_1 and u_2 (cf. Fig. 1.4). (b) describes the landscape in two dimensions, as in a contour map.

$$u_1 + u_2 = b \tag{1.6}$$

Equation 1.6 is a particular example of a trade-off since if u_1 increases u_2 must reduce by a specific amount and vice versa. In general a trade-off can occur along a curve, called a **trade-off curve**. The trade-off specified by equation 1.6 is plotted on axes u_1 and u_2 in Fig. 1.6a. However fitness contours can also be plotted on these axes (Fig. 1.6b) and when the two are superimposed (Fig. 1.6c) the optimal strategy can be found by inspection; it is the point (i.e. combination of u_1 and u_2) on the trade-off curve (i.e. a physiologically feasible combination of u_1 and u_2) which is coincident with the highest fitness contour (i.e. the physiologically-feasible combination that gives the highest fitness value).

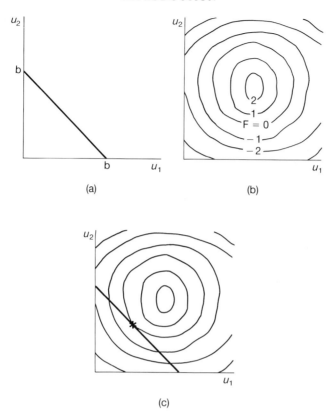

Fig. 1.6. (a) A possible trade-off between u_1 and u_2 (described by equation 1.6). (b) Fitness contours as in Fig. 1.5b. (c) Superimposing (a) and (b) allows us to find the optimal strategy (starred) by inspection.

This simple procedure provides a robust, quick and easy way of finding the optimal strategy and, where it can be applied, it is the method we prefer.

As can be seen in Fig. 1.6c the fitness contours are parallel to the trade-off curve in the immediate vicinity of the optimal strategy (this is intuitively obvious — as described above — but a general proof can be formulated using calculus), i.e. the slope of the trade-off curve is identical to the slope of the fitness contours, and this fact can sometimes be used to calculate the optimal strategy. The method consists of (1) finding an equation specifying the slope of the trade-off curve; (2) finding an equation specifying the slope of the fitness contours; (3) putting the two equations together. When the slopes are identical (which is characteristic of optimal strategies, see above) they can be

BOX 1.3. Finding (u_1^*, u_2^*), the optimal strategy in a two-dimensional case.

We apply steps (1)–(3) suggested in the text to a hypothetical example.

1 Suppose the trade-off curve is given by $u_1 + u_2 = b$ (as in equation (1.6) illustrated in Fig. 1.6a). Differentiating with respect to u_1 gives

$$\frac{\partial u_2}{\partial u_1} - 1 \qquad (1.7)$$

2 Suppose the fitness contours are circular with equation $(u_1 - 1)^2 + (u_2 - 1)^2 = (-F)^2$, where F is constant along a particular contour. Differentiating with respect to u_1, while holding F constant,

$$2(u_1 - 1) + 2(u_2 - 1)\frac{\partial u_2}{\partial u_1} = 0$$

$$\therefore \frac{\partial u_2}{\partial u_1} = -\frac{u_1 - 1}{u_2 - 1} \qquad (1.8)$$

This gives the slope of the fitness contours.

3 We can now find the optimal strategy (u_1^*, u_2^*), which is characterized by the slopes of the trade-off curve and the fitness contours being equal (see text). Thus $\frac{\partial u_2}{\partial u_1}$ can be eliminated between equations (1.7) and (1.8) to give

$$-\frac{u_1^* - 1}{u_2^* - 1} = -1, \text{ i.e. } u_1^* = u_2^*$$

But $u_1^* + u_2^* = b$ from equation (1.6) so

$$u_1^* = u_2^* = b/2 \qquad (1.9)$$

This is the optimal strategy.

eliminated between the two equations to give a single equation whose solution is the optimal strategy. An example of this method of finding the optimal strategy is given in Box 1.3.

The main advantage of using calculus to find the optimal strategy — as in Box 1.3 — is that the optimal strategy is then characterized by an equation which shows exactly which environmental parameters affect it, and specifies what the effects are. This method will later allow us to compare and contrast habitats, and to see what strategy is optimal in each (Chapter 7). However the calculus method is of only limited value if the optimal strategy involves extreme values of the us (so that some of them are 0) or if the various curves are not smooth, since in these cases one does not always know what the slope of the curve is.

It is now necessary to introduce a feature which improves the biological realism of these models but makes their analysis far more complicated. Assume that the allocation of resources to growth, u, varies with age, t, then the optimal solution is that function of time $u(t)$ which maximizes fitness. Under these circumstances there is another factor to consider — body size — which has implications for reproduction and survival. Let body mass be m. By allocating resources to growth, body mass is proportionately increased, i.e.

$$\frac{dm}{dt} = cu(t) \tag{1.10}$$

where c is a constant of proportionality. $u^*(t)$ can be found using a method of calculus known as **Pontryagin's Maximum Principle** (Box 1.4).

We refer to models summarized in Boxes 1.2 and 1.3 and the selective landscapes (Fig. 1.5) as *static* and those in Box 1.4 as *dynamic*. Sometimes the static models are appropriate; for example, in analysing patterns of allocation at particular points in life cycles when it can be assumed that size changes are minimal. This approach will be used in analysing allocation patterns in reproductive adults (Chapter 4). However, the dynamic models are usually more realistic and have to be used in the growth phase of the life cycle (Chapter 5). Moreover, throughout the book we shall be moving towards a general class of dynamic allocation models that attempts to summarize the complete complexity of physiological allocation patterns. Despite their complications, however, these will be based on the principles of allocation illustrated in Figs 1.2 and 1.4 and will always be referable to the basic

BOX 1.4. Finding u*(t), the function of time which maximizes fitness.

It has been shown (Taylor *et al.* 1974, Sibly *et al.* 1985) that life cycles (i.e. S_t and n_t schedules) which maximize F in equation 1.4 can be found by the following procedure. Find the life cycle that maximizes the function

$$\Phi = \int_0^\infty e^{-Rt}\, S_t n_t\, d_t \qquad (1.11)$$

with respect to S_t and n_t where R is an arbitrary constant. Do this for a range of different values of R and find the one for which the maximum value of Φ is 1. The number R and the associated pattern of resource allocation are respectively the maximum attainable value of F and the corresponding optimal strategy. This provides the basis of a method for finding the optimal strategy, by optimizing Φ in equation (1.11). To do this we use Pontryagin's Maximum Principle to solve the problem of finding a vector variable $\mathbf{u}(t)$ (the optimal strategy) for all values of t from initial time 0 to terminal time T so as to maximize the function

$$\Phi = \int_0^T f(\mathbf{x}(t),\, \mathbf{u}(t))\ dt \qquad (1.12)$$

where

$$\frac{d\mathbf{x}}{dt} = g(\mathbf{x}(t),\, \mathbf{u}(t)) \qquad (1.13)$$

and $\mathbf{x}(0)$ is fixed. In addition *either* T (as on page 106) *or* $\mathbf{x}(T)$ (as on page 96) is fixed. Equation (1.13) is referred to as the 'state equation' and relates $\mathbf{x}$, referred to as the state variable, to control variable $\mathbf{u}(t)$ — an example of a state equation is equation (1.10). One of the conditions that must be satisfied by the optimal strategy is that for every value of t from 0 to T, $\mathbf{u}(t)$ must have the value that maximizes the 'Hamiltonian', H:

$$H = f(\mathbf{x}(t),\, \mathbf{u}(t)) + \lambda(t) \cdot g(\mathbf{x}(t),\, \mathbf{u}(t)) \quad (1.14)$$

in which $\lambda(t)$ is a function of t such that

$$\frac{d\lambda}{dt} = -\frac{\partial H}{\partial \mathbf{x}} \qquad (1.15)$$

If T is fixed and $\mathbf{x}(T)$ is free then

$$\lambda(T) = 0 \qquad (1.16)$$

Alternatively if T is free and $\mathbf{x}(T)$ is fixed then

$$H(T) = 0. \qquad (1.17)$$

We try to give a biological interpretation of the λs in Chapter 8, but here we offer a provisional, non-rigorous explanation, based on Alexander (1981). To increase fitness, it is desirable to reproduce fast, but it is also desirable to grow fast so as to be bigger (and able to reproduce faster) in the future. The Hamiltonian (equation 1.14) has two terms, one (**f**) representing reproductive rate and the other (**g**) growth rate. Lambda is a weighting factor that expresses the relative importance of the two terms. Early in life, growth may be advantageous. As the end of reproductive life approaches, growth is less useful, because there will be little time to take advantage of increased size. Therefore lambda approaches zero with increasing age.

principles of optimization theory that are illustrated by example in Boxes 1.2–1.4.

1.6 The principle of compensation

In Chapters 4 and 7 we will make comparisons between habitats in which different selection pressures operate. Comparisons of this sort involve a potential pitfall. Fortunately this is not very serious, but it needs to be kept in mind.

Imagine two stable populations (A and B) living in the habitats being compared. Suppose that mortality rates are higher in environment A, but that birth rates are the same in both populations. Since the populations being compared are assumed to be stable, it follows that (individual) growth rate must be higher in environment A, to compensate for the higher mortality rate, since otherwise with worse mortality and worse growth, population A must necessarily decline compared with population B. Thus if one population has a higher mortality rate it must have a higher individual growth rate (or birth rate) to compensate. The general principle is that mortality, growth and birth rates cannot vary independently of one another between stable populations. We shall refer to this as the **Principle of Compensation**.

Note that the emphasis is on comparison between, not within, populations. We are not at this stage making any claims about population regulation or the factors which make a population stable. Instead we are making a comparison between two stable populations. For simplicity the principle has been stated for stable populations ($F = 0$) but it could, if required, be extended to the case where the populations being compared have the same non-zero F value.

So far we have said nothing about how compensation might come about in practice. However, the mechanism might be ecological regulatory factors that operate in a density-dependent fashion. Consider, for example, the sequence of events following an increase in the mortality rate in a population which had previously been stable ($F = 0$). The increase in mortality rate necessarily produces population decline which, if unchecked, would lead to extinction. However as the population reduces and population density decreases more food might become available to juveniles so that they breed much earlier, with the result that the population achieves a new steady state. Note that this is only an example and that in reality the compensating factors might be any combination of growth rates, mortality rates and fecundity.

With this sort of ecological mechanism in mind we can illustrate the possible pitfalls in comparing habitats. Consider the following example which we will discuss further below (p. 131). An investment in gonads will lead to an enhanced fecundity, but during the period of preparation it might lead to an enhanced pre-reproductive mortality risk for the parent — e.g. courtship, copulation, pregnancy and parturition are all risky. If extrinsic risks for the parent are increased, e.g. due to increased predation, *it could be optimal to reduce the investment in the gonads*, and hence n, otherwise by combining a high extrinsic with a high intrinsic mortality (due to reproduction) the parent might not survive to reproduce at all. However increased extrinsic adult mortality risk, might mean adults, as a class, ate less so that there was more food available in the habitat for juveniles, as a class, and hence growth rates and survival rates for juveniles would improve. This could be the basis for the ecological compensation that leads to a new population equilibrium. But increased juvenile viability could mean that the risks associated with preparing for reproduction are worth taking so it would be optimum to enhance not reduce investment in gonads.

Hence the Principle of Compensation can complicate the optimality argument. Remember, though, this is true only if the fitness components that undergo compensation are involved in the optimality arguments, which is often not the case; i.e. there are often sufficient **degrees of freedom** in the system to allow for compensation without modification of the optimum. For example the hypothetical animals shown in Fig. 1.6 might profit from a beneficial mutation affecting u_1 or u_2 which increased the fitness of the animals carrying it, so that the mutation spreads to fixation in the population. If compensation involved modification of variables other than u_1 and or u_2 and did not affect the shape of the fitness contours then the optimality analysis shown in Fig. 1.6 would be unaffected by compensation. Moreover, it often turns out that different fitness components have opposite effects on the optimum. Thus, an enhanced juvenile survival, but a reduced adult post-reproductive survival, should favour enhanced investment in reproduction. A compensation in the survival of one age class can only occur in the opposite direction to which the other is perturbed — so compensation would here accentuate what is predicted by the optimization argument. Hence in what follows, we shall generally not need to bother with the Principle of Compensation, but we shall point out places where it may complicate the predictions from the models.

Furthermore, in many of the chapters we are not making comparisons between habitats. Instead we are concerned with defining general factors that enhance fitness — i.e. would allow the invasion of a population by a new mutation — and here the Principle of Compensation is not relevant at all (i.e. Chapters 2, 3, 5, 6 and 8).

1.7 Principle of allocation reconsidered

As we have emphasized at several points above, the Principle of Allocation is one important basis of physiological trade-offs. So long as it is applied to what an individual organism can do with resources from a particular meal or over a particular time (e.g. a lifetime) then, because of the conservation laws, it is irrefutable. It also applies to different organisms with the same resource input and in applying it we often assume that this and all other aspects of metabolism are held constant, except for shifts in the allocation of resources between specified physiological demands; in Fig. 1.4 'Food' is constant but u_1 and u_2 can change. It is as if we introduce different gene-determined resource allocation strategies into the same physiological background in order to judge their fitness.

The possibility of different levels of resource input, e.g. through different rates of feeding, threatens the Principle of Allocation (Tuomi *et al.*, 1983) but not necessarily optimization philosophy. Organisms with different levels of resource input, still have to operate within the resource limits at these levels. Hence a trade-off operates at each level but optima shift between levels according to resource supply. In other words, resource input becomes an extra dimension in optimization models (Fig. 1.7). Specific examples of this will be given in Chapters 3, 4 and 7.

Finally, it should be noted that physiological constraints and trade-offs do not only emanate from the Principle of Allocation. For example, there can be negative correlations between physiological rates and efficiencies and models based on these will be discussed in Chapters 2 and 3. Similarly, there can also be negative correlations between physiological rates and survival chances, some but not all of which depend on the Principle of Allocation, and examples will be considered further in Chapters 3, 4, 5, 7, and 8.

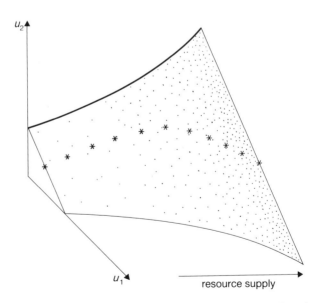

Fig. 1.7. The extra dimension of resource supply added to Fig. 1.6a. Putative optima are indicated by stars.

1.8 Currency

Another important consideration in using the 'economics approach' is in choosing a currency. So far we have referred rather vaguely to 'resource'. More formally, this could be treated as a vector quantity with each variable representing a subcomponent of food or tissue; e.g. proteins, fats, carbohydrates, vitamins, or carbon, nitrogen, phosphorus, or energy and matter. In principle, selection will have most impact on those variables that are required most widely in metabolism and which are most limiting. Because of the limiting nature of the resource-acquiring processes it can be argued that all resources are likely to be more or less limited (p. 10). Hence, generality becomes an important criterion and the one resource which is required most generally, indeed universally, in both processes and structures, is energy. This is needed to mobilize and power the use of all other resources. Gutshick (1981) considers energetics to be at the heart of nitrogen budgeting in plants but Abrahamson & Caswell (1982) found poor correlations between biomass and mineral allocation patterns. For animals, energy is likely to be a good general currency when food and feeder are of similar biochemical composition (i.e. for carnivores), but less good for herbivores and detritivores which often live in a 'sea' of carbohydrate (energy) but face shortages of protein, nitrogen, vitamins and certain trace elements. Thus although in particular situations energy may not be the most limiting and hence most suitable currency, nevertheless it will be the most generally suitable currency. Moreover, it is relatively easy to measure and also manipulate in theoretical models. Hence, whenever it becomes necessary to discuss models more particularly than in terms of 'resources' we shall usually use energy as our currency.

Another note of caution is also appropriate at this point on the general validity of the economics approach. An economics of metabolism with energy as currency emphasizes the quantitative aspects of the phenotype. However, the quality of behaviour and form will often be important in determining survival and reproductive success. For example, the organization of a nervous system is probably as important to the way it works as its size. Size, nevertheless, puts important constraints on form and organization even in nervous systems (see Lewin (1982) reporting on the work of R.D. Martin 1981). And, as we shall argue below (Chapter 8), organization and form are generated by the differential allocation of resources between structures in space. Hence, the economics of resource use can make important statements about form and organization and then about the consequences of this form and organization.

Table 1.3. General scheme of the book

Chapter	Subject	Measure of fitness (Φ)	Type of model	Basis of trade-off
2	Acquisition	auxiliary (phenotypic)	static	rates vs. efficiencies
3	Metabolic costs	"	"	" " Principle of Allocation
4	Reproduction	direct (F)	"	Principle of Allocation
5	Somatic production	direct (H)	dynamic	rates vs. risks (possibly related to Principle of Allocation see Chapter 3)
6	Size	direct (F)	static	
7	Habitat classification		based largely on Chapter 4	
8	Allocation in general	direct (H)	dynamic	general

1.9   Conclusions and prospects

Concepts of physiology and evolutionary biology can be combined in models based on the assumption that organisms are economic units, optimized by natural selection, subject to certain physiological trade-offs and constraints. The approach is consistent with a hypothetico-deductive philosophy in which models are constructed *a priori* and predictions from them tested in carefully controlled experimental situations. Two challenges for physiology emerge from these models; one in identifying the trade-offs and constraints to be incorporated into them and the other in testing their predictions. Physiology, therefore, not only has something to learn from evolutionary biology (in the predictions) but something to contribute to understanding how natural selection works (the trade-offs and constraints). This book focuses on the theoretical tools through which these enrichments of physiology by evolutionary biology can be achieved. We begin with a chapter on resource acquisition, which interfaces between behaviour (selecting and getting food) and physiology (digesting and absorbing it), using as phenotypic measures of fitness either the net rate of food intake, or the net rate of energy supply from the gut to the body, as appropriate. In Chapter 3 we deal with allocation of resources between maintenance, activity, and the production of new tissue, and here the production of tissue is the appropriate fitness measure. These fitness measures will be satisfactory so long as mortality and other costs are held constant, i.e. do not vary between phenotypes, since increasing the rate of energy supply or tissue production increases fecundity and growth rate and allows breeding to occur sooner, as discussed above. Mortality factors and trade-offs are introduced at the end of Chapter 3 and in Chapter 4.

In developing the scope of our analysis (summarized in Table 1.3) we are guided partly by mathematical rather than biological logic, since we wish to progress from simple static models based on auxiliary hypotheses to more complex static models based on direct measures of fitness and finally to complex dynamic models. In so doing we have no hesitation in treating reproductive production before somatic production, even though, biologically, the former must precede the latter, because we assume that most readers will be more comfortable with the biology than with the mathematics. At this point readers might like to skip to the general Summary (p. 157) to get an overview of the book's contents and structure.

2
Feeding and Digestion

2.1 Scope

Food items are acquired and processed to supply nutrients to the body. We shall deal only briefly with the question of food acquisition because this is at the interface of behavioural and physiological ecology and has been dealt with extensively before (Krebs & McCleery, 1984). Instead we shall focus more attention on how food items should be processed during digestion, for this is more properly the domain of physiological ecology. The approach will be *a priori* but the models will be based on auxiliary hypotheses and in particular on the assumption that the net rate of obtaining energy is a suitable 'phenotypic measure of fitness' (Chapter 1). Foraging and digestive strategies that maximize this criterion should therefore be favoured, and this yields predictions that can be tested by observation and experiment.

2.2 The theory of optimal foraging

In this theory the world is divided conceptually into patches, each containing food. These could be spatially separate areas in which food occurs, for example mussel beds visited by shore birds, clover fields visited by woodpigeons, trees with berries visited by song birds, flower clusters visited by insects, and so on. Optimal foraging theory is concerned with the order in which patches should be visited, and the length of time that should be spent in each patch to maximize the net rate of obtaining energy, bearing in mind that time and energy are expended exploiting and travelling between patches. The optimal strategy, using the criterion defined above, is characterized as follows:

A feeder should exploit patches which initially yield energy at a net rate above or equal to that obtained in the environment overall. If the yield drops to a level equal to that which could be obtained by the feeder in the environment overall, it should leave the patch. To put it another way, a feeder should not stay in a patch when it could do better by travelling to another one. Part of this argument is formalized in Box 2.1.

Patches that initially yield less than the overall rate of obtaining energy should not be exploited at all. Thus if a forager, selectively foraging in good patches, happened to encounter a poor patch it should feed there if and only if the initial yield was greater than the overall rate at which it had been previously obtaining energy.

By a simple modification, the above theory can be extended to make predictions about optimal diet. Consider what happens as individual patches become smaller and smaller. In the limit they become

---

BOX 2.1. Calculation of the optimal time to spend in a patch if the travel time (between patches) is fixed.

The curve in Fig. 2.1 represents the energy obtained from a patch in relation to the time spent exploiting it (i.e. a pay-off curve). The patch illustrated yields returns at a diminishing rate. The slope of the line A to B is equal to (energy obtained by spending time t_B in each patch)/(travel time + t_B) i.e. rate of obtaining energy. Hence rate of obtaining energy is maximized by the steepest line from A to the curve (starred).

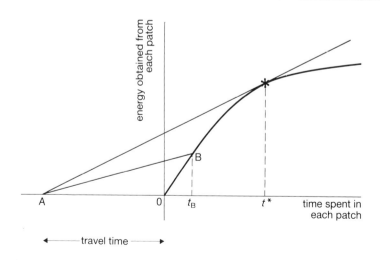

Fig. 2.1. Calculation of the optimal time t* to spend in each patch (see Box 2.1).

equivalent to individual food items. The theory can now make predictions about which food items should be consumed in order to maximize the net rate of obtaining energy, again bearing in mind that time and energy are expended handling and travelling between food items. The optimal diet is characterized as follows:

A feeder should exploit food items which yield energy at a net rate (after allowing for handling time) above or equal to that obtained in the environment overall. If, on inspection, a food item seems inferior the feeder should discard it, if eating it lowers the overall net rate of obtaining energy. To put it another way, the feeder should not eat a food item if it could do better by looking for another one. This argument is formalized in Box 2.2.

The above gives a simple outline of optimal foraging theory in terms which are readily formalized into simple mathematical models (Boxes 2.1 and 2.2). Predictions derived from these models have been extensively tested both in field and laboratory and the evidence is

---

BOX 2.2. Calculation of the optimal diet.

Suppose an animal is eating all the best food items, encountered every T_0 secs and yielding E_1 energy units if processed for T_1 secs, so that the rate of obtaining energy is $\dfrac{E_1}{T_0 + T_1}$. Other food items yielding E_2 energy units if processed for T_2 secs should be eaten only if

$$\frac{E_2}{T_2} > \frac{E_1}{T_0 + T_1}$$

---

generally supportive, though the tests are often limited by difficulties associated with accurately measuring energy yield and energy expenditure (Krebs & McCleery, 1984).

One way the above theory has been extended is by contrasting the optimal strategy for using patches in a good environment (yielding a high overall rate of energy intake) with the optimal strategy for using patches of the same type in an environment which is poorer (because the patches are further apart so that more time is spent travelling between them). The optimal way to forage in any given environment is to leave each patch when the rate of obtaining energy from it drops to the level obtained in that environment overall. However this will be higher in the better environment, and so animals in the better environment should not stay so long in individual patches. Thus if travel time between patches is decreased, individual patches should be exploited less. This prediction has been tested by Cowie (1977) in an experiment using captive great tits which searched for worms hidden in patches (sawdust-filled cups). When the travel time was increased the birds spent longer in each patch before leaving. Moreover there was a convincing fit between the results and the predictions. However in an extremely poor environment it is possible that animals should attempt to leave the area in the hope of finding better foraging elsewhere.

2.3   Constraints imposed by food items

The situation is made more complicated by the various defensive and incentive properties of live food items. These have themselves been subject to selection, usually to reduce the amount of damage caused by animals foraging, but sometimes to exploit animal foraging, as in the provision by plants of incentives to foraging animals in the form of nectar or fruit which improve the dispersal chances of pollen and seeds (Enckell & Nilsson, 1985). When plants contain deterrents in their leaves and structural supports, this places constraints on the foraging strategies of foliovores, which are then limited in their intake of that species by their ability to cope with the plant deterrents. Examples are the defensive chemicals, including tannins and possibly cellulose, in plant leaves. Janzen (1981) argues that there are many structural polysaccharides other than cellulose which could be used as the main constructional material in plants, that cellulose is relatively expensive because it limits a plant's repair options and it cannot be recovered from leaves which are shed, but that the price is worth paying because virtually no higher organism can digest cellulose unaided. Of course herbivores can respond to these defences by evolving their own defence systems. For example, mixed-function oxidase (MFO) enzymes are widespread in insect herbivores (e.g. Brattsten et al. 1977) and are capable of degrading some of the toxins produced by plants to a form that is easily metabolized. In turn these MFOs might impose a 'metabolic load' in the production and maintenance of detoxifying systems. However, from a comparison of the growth of several species of lepidopteran caterpillar — some generalist, others specialist feeders — Scriber & Feeny (1979) conclude that the 'metabolic load' has only a trivial effect on larval growth rates.

Other complications arise because animals require other nutrients as well as energy, so the net rate of obtaining energy may not be a satisfactory phenotypic measure of fitness. One possibility is to consider the nutrient requirements as a constraint, so the optimization of energy intake is subject to some minimum amount of the nutrient being obtained. This approach has been used by Belovsky (1978) in modelling the sodium requirements of moose (Fig. 2.2).

Further complications arise because when feeding, animals take mortality risks, which can be reduced by vigilance, or by feeding in groups or close to cover. The trade-off between food intake and mortality risk has received only a limited amount of attention (Krebs & McCleery, 1984; Caraco *et al.* 1980; Sibly & McFarland, 1976).

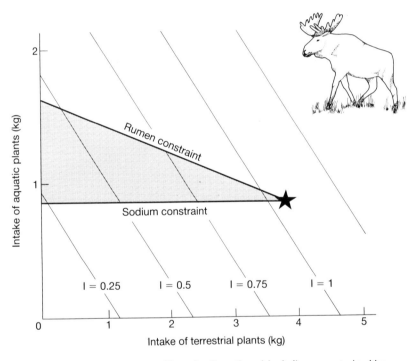

Fig. 2.2. Optimal diet for moose. The animal's options (shaded) are constrained by rumen capacity (the 'rumen constraint') and by its daily need for sodium (the 'sodium constraint') which is mainly found in aquatic plants. Also shown are lines of equal energy intake (*I*) in multiples of metabolic rate, and the optimal diet (starred) corresponds to the highest *I* allowed by the model. Moose actually eat the mixture of plants predicted by this model. After Belovsky (1978).

2.4   The theory of optimal digestion

Once obtained by the animal, food is processed through a set of chambers illustrated schematically in Fig. 2.3a. Unfortunately there are few data on the quantitative relationship between the amount absorbed from a meal and the time that it is retained in the gut, but it seems likely that the relationship illustrated in Fig. 2.3b for an octopus, where the cumulative amount extracted increases, but at a reducing rate, with retention time, is a general one. The amount of resource extracted per unit of food ingested is referred to as **digestive efficiency**, and therefore Fig. 2.3b can be thought of as showing how digestive efficiency can be improved by increasing retention time. A better measure of digestive efficiency, however, would also take into account the energy and other resources invested in processing the food, so that the y axis would represent the net energy obtained from one gram of food. This is shown in Fig. 2.4 which represents the progress of digestion as a graph in which the net energy obtained from one gram of food, k_d, is plotted against the time the digesta have been in the gut (retention time, t_d). During the first phase of digestion, time and energy are invested in breaking down the food's defences, leading to a decline in the net amount of energy obtained. When these defences are breached and the food cells have been separated into relatively small molecules, these small units are rapidly absorbed across the gut wall. Finally the rate of absorption declines as the gut completes the digestion of all that it is equipped to process (see Box 2.3 for a possible model of this process). If the digesta were retained beyond this point the cost of carrying them would lead to a decline in the net energy obtained.

The option open to an animal, given the shape of the digestion curve in Fig. 2.4, is to vary retention time. Next we consider what retention time would be optimal, i.e. what retention time would maximize the net rate of obtaining energy? We consider the simplest type of digestive system first.

2.5   Continuous flow digestive systems

The simplest digestive system is one in which food is continuously entering the digestive chambers at one end and undigested residues are constantly leaving from the other. This is the case for herbivores that graze continuously, such as horses or geese which eat grass. Suppose food items enter the digestive chambers at rate v (g/s) and are retained for time t_d before the undigested residues are evacuated. Then the

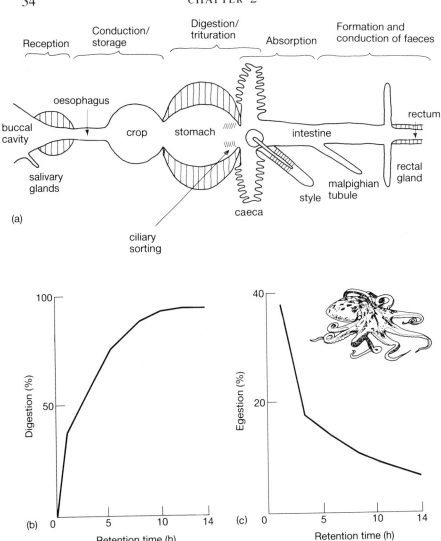

Fig. 2.3. (a) Schematic representation of a composite gut. No single animal has all these structures. Shaded walls represent muscles. (b) Cumulative uptake of food from a meal undergoing digestion in octopus gut (as a percentage of food ingested). (c) Simultaneous loss of faeces (per hour) as a percentage of food ingested. After Boucher-Rodoni (1973).

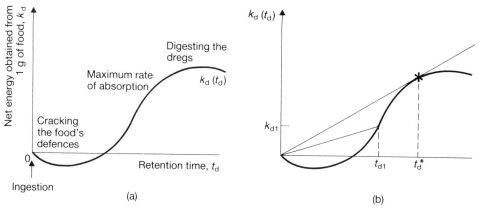

Fig. 2.4. Simple model of digestion. (a) The *net* amount of energy obtained may decline with time while the food's defences are being cracked because energy is provided to crack them but none is absorbed. When the defences are breached energy is rapidly absorbed, but eventually rate of absorption declines when no more food can be digested. $k_d(t_d)$ equals digestive efficiency plotted as a function of retention time. (b) The net rate of obtaining energy is $k_d(t_{d1})/t_{d1}$ if the retention time is t_{d1}, and this is the slope of the line from the origin to the curve $k_d(t_d)$. $k_d(t_d)/t_d$ is maximized by the steepest line from the origin to the curve (starred).

weight of material in the digestive chambers is approximately given by

$$\text{weight carried} \approx v \cdot t_d \qquad (2.1)$$

(see Sibly, 1981 for an exact expression). Since food enters the digestive chambers at rate v and each gram eventually yields k_d units of energy after retention for time t_d (Fig. 2.4), it follows that (net rate of obtaining energy, Js^{-1}) = (rate of ingesting food, gs^{-1}) × (energy obtained per gram ingested, Jg^{-1}),

$$\text{i.e. net rate of obtaining energy} = v \cdot k_d(t_d) \qquad (2.2)$$

Here $k_d(t_d)$ means a quantity k_d that is a function of t_d. From equation (2.1) v = (weight carried)/t_d and substituting this into equation (2.2) we have:

$$\text{net rate of obtaining energy} = (\text{weight carried}) . k_d(t_d)/t_d \qquad (2.3)$$

Therefore for a given weight carried, the optimal strategy is to maximize $k_d(t_d)/t_d$. This holds in particular when gut capacity is a limiting

BOX 2.3. Model of intestine operation.

Suppose that after intial processing, particles (of digestive products ready to be absorbed) of constant diameter d fill a cylindrically-shaped gut of diameter c,
length l, and surface area πcl. Approximately $\pi cl/d^2$
particles will be in contact with the gut wall (if the
particles are packed against the wall on a square lattice) out of a total of approximately $\pi c^2 l/4d^3$ particles
in the gut (if the particles are packed in the gut in cube
lattice formation). Thus the ratio of particles being
processed to particles present is $4d:c$. Suppose the gut
is randomly churned by muscular action every T
seconds, after all particles touching the gut wall have
been processed. Thus the chance per unit time of a
particle coming into contact with the gut wall is approximately $4d/cT$, and so the chance a particle has not
been processed after time t is approximately
$\exp(-4dt/cT)$, and the chance it has been processed is
$1 - \exp(-4dt/cT)$. Thus the fraction of particles processed after time t is $1 - \exp(-4dt/cT)$. This is a
negative exponential curve, and the speed of the
process depends on $4d/cT$. $4d/cT$ is maximized by maximizing d and minimizing c and T, but T is probably
limited by the way material is transported across the
gut wall, and d, although it depends on what the
animal was able to ingest, is probably also limited
below a critical size above which it becomes difficult to
transport particles across the gut wall. Reducing c
means that a greater surface area of gut has to be
maintained per unit gut capacity.

factor. Optimal retention time t_d^* can then be found graphically as in
Fig. 2.4b. It follows that animals eating poorer quality food should
have larger digestive chambers, other things being equal. This is because $k_d(t_d)/t_d$ is lower for a poorer quality food, and so $k_d(t_d^*)/t_d^*$ is
necessarily lower. But if the animal's nutritional requirements are the
same, and $k_d(t_d)/t_d$ is lower, the animal must carry a greater weight of

digesta (from equation (2.3)). This provides an explanation of the widespread finding that animals sustaining themselves on poorer food have larger guts. Moreover, there is some experimental evidence that on poorer diets some birds grow longer guts (Table 2.1). Acclimatization from rich to poor diet reverses the effect of acclimatization from poor to rich diet and takes about 25 days.

Table 2.1. Experimental evidence that birds on a poorer diet grow longer guts. After Sibly (1981)

Bird	Body weight (g)	Poor diet	Rich diet	Intestine length (cm) Poor	Intestine length (cm) Rich	Intestine length (cm) Poor/ Rich	Caecum length (cm) Poor	Caecum length (cm) Rich	Caecum length (cm) Poor/ Rich
Quail	125	artificial	artificial	51	46	1.1	17.0	14.5	1.2
Woodpigeon	430	brassica	grain	220	157	1.4	—	—	—
Starlings	75	plant	animal	33	27	1.2	0.8	0.6	1.3

2.6 Deposit feeders

Following pioneering models of filter feeding by Lehman (1976) and Lam & Frost (1976), Taghon et al. (1978) devised a model of optimal digestion in deposit feeders. The animals are assumed to feed continuously on organic material *on the surface* of ingested particles, and the gut is assumed to be always full. $k_d(t_d)$, the *net* amount of energy obtained from one gram of food retained in the gut for time t_d, is

(energy absorbed) − (energy invested in digestion), as in Fig. 2.4.

Energy absorbed as a function of retention time is assumed to be generally negative exponential in shape as shown in Fig. 2.5a. The energy invested in collecting and digesting particles when ingestion rate is v is assumed to be proportional to v^2. If the gut is operating at maximum capacity, then $t_d = $ (gut capacity)$/v$ (from equation 2.1) and so the energy costs of ingesting food at rate v are proportional to ((gut capacity)$/t_d)^2$. A hypothetical case is shown in Fig. 2.5a. Subtracting the energy costs of ingesting food at rate v from the energy absorbed gives $k_d(t_d)$. The optimal retention time (starred) is found by drawing a tangent from the origin to k_d (Fig. 2.5a).

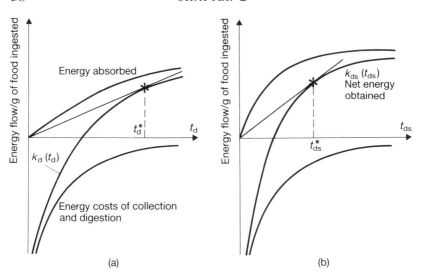

Fig. 2.5. (a) Net energy obtained from 1 gram of food by deposit feeders $k_d(t_d)$ = energy absorbed − energy cost of collecting and digesting the food particles. (b) If the animal were ingesting smaller particles (better quality food) energy would be absorbed faster and the optimal retention time t_{ds}^* would be shorter than in (a).

What are the implications of changing the particle size? Taghon *et al.* (1978) assumed that the digestible material consisted of microbes growing on the surface of ingested particles. Hence the quantity of this material was presumed to be proportional to the surface area of the particles (NB this differs from Box 2.3, where the particles themselves were assumed to be absorbed across the gut wall). Thus smaller particles, which offer greater surface area per unit volume, represent better quality food, e.g. halving the diameter of the particles should double their combined surface area. This situation is represented in Fig. 2.5b. Note that the optimal retention time is shorter but that the rate of obtaining energy is greater if the particles are smaller. (These results do not depend upon the precise form of the energy costs of ingesting food at rate v, provided it is a monotonic decreasing function of t_d, as in Fig. 2.5). Hence one would expect deposit feeders to select smaller particles and to process them faster.

Taghon (1982) designed elegant field experiments specifically to test whether smaller particles are in fact selected, using two sizes of artificial particles and a range of animal species. He found that most species did select the smaller particles, as predicted (Fig. 2.6). However in every case large particles were also taken, possibly because of

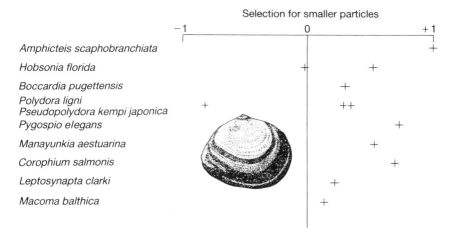

Selection for smaller particles

	−1	0	+1

Amphicteis scaphobranchiata

Hobsonia florida

Boccardia pugettensis
Polydora ligni
Pseudopolydora kempi japonica
Pygospio elegans

Manayunkia aestuarina

Corophium salmonis

Leptosynapta clarki

Macoma balthica

Fig. 2.6. Deposit feeders generally select smaller particles when offered a choice in feeding experiments. Selection for smaller particles is defined as (proportion of smaller particles in guts)/(proportion of smaller particles in sediment) −1. On this graph a species value to the right of the mid line indicates that the species selected smaller particles. Data from Taghon (1982).

limitations in handling the particles imposed by the use of mucus to pick up smaller particles selectively.

2.7   Complications in digestion

So far we have only considered the simplest type of digestive system in which food is continuously and steadily passed through the digestive chambers. However if food is distributed in patches (Section 2.2) there will necessarily be periods when the animal is travelling between patches and is therefore unable to eat. What effect will this have on digestion? One possibility is for the residue from one patch to be evacuated when the next patch is encountered, but this yields suboptimal energy re-turns, according to Fig. 2.4, except in the unlikely event that the time taken travelling between patches is equal to the optimal retention time. Therefore if the next patch was encountered before optimal retention time was reached, then the animal should start eating imme-diately without evacuating residues from the previous patch. However if the next patch was not encountered until after optimal retention time had passed, then there might be a period after defaecation when the gut was empty. The optimal strategy in this case is to continue digestion until digestive efficiency reaches its maximum value (i.e. the $k_d(t_d)$ curve reaches its peak, Fig. 2.7) and then evacuate the gut.

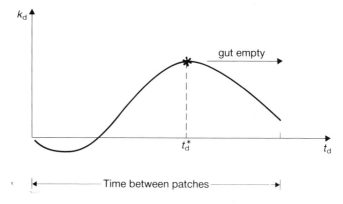

Fig. 2.7. Optimal digestion when time between patches is large. The optimal strategy (*) is to evacuate the gut when $k_d(t_d)$ reaches its peak.

An alternative would be to store food (for example in a crop) until space becomes available to digest it, but this increases the weight carried, and necessitates the maintenance of extra tissue which is not always in use. Whether these costs are worth paying depends on whether they produce worthwhile benefits (see Chapter 3).

Animals subject to relatively long gaps between meals generally digest food quickly immediately after the meal but more slowly when little remains (Fig. 2.3b). At first sight this seems paradoxical — if there were no cost associated with processing food quickly then the animal should process it all at the maximum possible rate. Then un-digested food would not be carried around longer than necessary, the digestive process (involving, e.g. blood supply to the gut) would take the shortest possible time, and energy would be extracted from the food as quickly as possible. Because the gut does not operate in this way it seems likely that there is some cost associated with processing food at the maximum possible rate. Some possible costs are:

1 less can be extracted from food processed faster;
2 the gut may suffer more wear and tear;
3 the other processing costs (blood supply, enzyme production) might increase;
4 the processing facilities (e.g. enzymes) supplied at the start might be used most efficiently by a diminishing rate of digestion.

(See Sibly, 1981, for further discussion.)

Although these possibilities are not mutually exclusive, the first is the most attractive. It fits with our understanding of digestion summarized in Fig. 2.4a where it was suggested that more energy is obtained from digesta retained for longer. To put it another way, digestive efficiency should decline as the throughput of digesta increases. In a useful review, Lawton (1970) concluded that generally the same or less is extracted if the feeding rate is higher (Table 2.2).

Because the constituents of food are not all equally accessible during digestion it is worth examining the possibility of sorting components of digesta into different types of compartment (e.g. caeca) in which independent digestive processes occur. For example, in rabbits coarse particles pass straight down the gut and their remains are egested relatively quickly as hard faecal pellets. Smaller particles, however, are retained in a long caecum with a well-developed vermiform appendix which harbours carbohydrate-digesting bacteria. After processing they are egested as soft pellets and immediately re-ingested. The soft pellets lie intact in the fundus of the stomach for at least six hours while bacterial fermentation continues, thereafter their processed contents are absorbed from the small intestine in the usual way (Lang, 1981). Thus coarse and fine particles are subjected to different digestive processes, with different retention times. The advantage of this system can be seen from Fig. 2.8 in which two different digestive processes are subject to different digestion functions $k_{d1}(t_{d1})$ and $k_{d2}(t_{d2})$. By operating the two processes separately, each can be optimized independently, necessarily producing returns greater than, or

Table 2.2. Review of relationships between digestive efficiency and feeding rate

Type*	Relationship	Examples
A	digestive efficiency remains constant over a wide range of feeding rates	the damselfly, *Pyrrhosoma*; many crustaceans; the sunfish, *Lepomis*
B	digestive efficiency decreases with increasing feeding rate	many fish; the terrestrial isopod, *Armadillidium*; *Clava, Daphnia, Artemia*
C	digestive efficiency increases with decreasing feeding rate	rare; the goldfish, *Carassius*

* After Lawton, 1970.

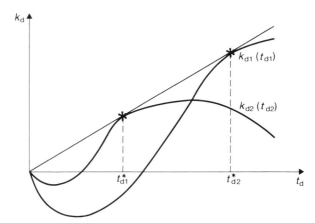

Fig. 2.8. Two food constitutents, subject to digestion curves $k_{d1}(t_{d1})$ and $k_{d2}(t_{d2})$, can be independently optimized if they are processed separately. The optima must share a tangent through the origin, since if one tangent were shallower than the other, the food to which it relates would not be worth eating (this argument does not apply if the processes extract different nutrients).

equal to, those obtainable within a single compartment. This shows the benefit that could accrue through the use of caeca. Of course the costs of maintenance of such extra organs should also be taken into account, using the methods described in the next chapter.

2.8   Conclusions and summary

The auxiliary, *a priori* hypothesis that foraging and digestive strategies should maximize net energy returns can give considerable insight into the evolution of feeding processes and structures. The rule for foraging is that feeders should not stay in a patch or eat a food item if they could do better feeding elsewhere; to be more precise, they should leave a patch or ignore a food item if the net rate of energy returns from it is less than or equal to returns from the environment in general. Once food is obtained, the energy returns from a meal are likely to vary non-linearly with the time that the meal is retained in the gut and on this basis it is possible to explain why continuous feeders often have larger guts when feeding on poor quality food, and why deposit feeders selectively feed from small rather than large sediment particles. The fact that discontinuous feeding animals often digest food quickly after a meal and then more slowly as the post-meal time increases is more of a puzzle. The most reasonable hypothesis, though, is that rapid digestion carries a cost, most probably in terms of impaired digestion. Compartmentalization of the gut can be an advantage because processes operating in different compartments, on different fractions of the food with different properties, can be optimized separately.

3

Costs of Living

3.1 What are they?

Physiological processes that have positive effects on fitness, via survivorship, S, and reproductive rate, n, at age t, (Chapter 1), all require power. This is usually supplied as ATP generated from the controlled catabolism of food resources and/or body stores, in a process familiarly known as respiration. Most of the energy allocated to respiration is ultimately dissipated as heat. We refer to these energy losses as **costs of living** and they are summarized schematically in Fig. 3.1. In this chapter we give consideration to three important components of the costs of living: costs of activity (in obtaining food); costs of tissue synthesis (influencing growth and hence t and also n); and costs of maintenance (influencing the viability of organisms and thus their survival chances, S). Costs of physical activity will be considered in relation to their effects on the production of synthesized tissue, P. Activity also influences survival probabilities and these benefits can be related to the level of energy investment in activity on the assumption that the relationship between activity and survival is optimal in the environment in which the animal lives. Again the models are based largely on the auxiliary phenotypic fitness measures — to maximize the net energy returns — discussed in the last chapter.

3.2 Respiration

The mechanisms involved in generating ATP are surprisingly uniform throughout the animal kingdom and are summarized in Fig. 3.2; for more details the reader should refer to Lehninger (1973).

The primitive system operated in an atmosphere without oxygen and may have been equivalent to the glycolytic–lactate pathway found in many extant organisms. It can be summarized as:

$$\text{Glucose} + 2\ \text{ADP} + 2P_i \rightarrow 2\ \text{Lactate} + 2\ \text{ATP} + 2H_2O$$

44

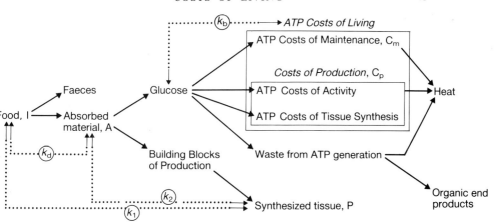

Fig. 3.1. Energy flows per unit time (solid arrows), and the efficiencies ($\leftarrow \cdots O \cdots \rightarrow$) of the components of the system.

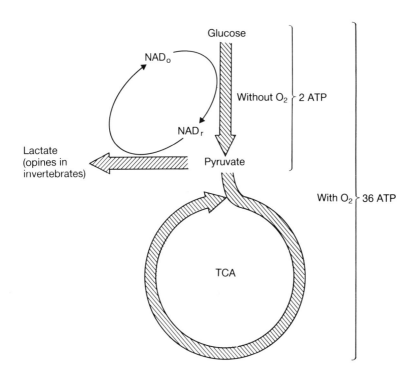

Fig. 3.2. An extremely stylized representation of the major metabolic steps concerned with the generation of ATP. From Calow & Townsend (1981b).

The ATP is generated from ADP and inorganic phosphate (P_i) by the direct involvement of substrates passing down the metabolic pathway; i.e. substrate-level phosphorylation. Related pathways, common in invertebrates, use the same glycolytic steps but instead of ending with the reduction of pyruvate to lactate they end with the condensation of pyruvate with an amino acid derivative to form a so-called opine.

With the evolution of photosynthesis and the accumulation of oxygen in the atmosphere of this planet, the aerobic component of the metabolic pathway became possible and this is known as the tricarboxylic acid (TCA) cycle; the overall reaction can be summarized as:

$$\text{Glucose} + 36\ \text{ADP} + 36\ P_i + 6\ O_2 \rightarrow 6\ CO_2 + 6\ H_2O + 36\ \text{ATP}$$

The TCA cycle generates reduced nicotinamide adenine dinucleotide (NAD_r), which is then used to convert ADP + P_i to ATP by donating electrons (and thereby becoming oxidized to NAD_o) to an electron transport system in which the final electron acceptor is oxygen. This is electron transport phosphorylation.

It is interesting to note that many invertebrates that experience long-term or continuous anoxia use alternatives to the glycolytic pathway in which the NAD_r, generated in the pathway from glucose to pyruvate, is not oxidized by the conversion of pyruvate to lactate but by the involvement of an electron transport system using an organic molecule, fumarate, instead of oxygen as the final electron acceptor. This increases the yield of ATP per glucose-unit input from 2 to between 4 and 6. Mechanisms of this kind, which we shall call fumarate systems, have been discovered in endoparasites and some marine animals that suffer long-term anoxia (Livingstone, 1983).

Thus there are major differences between the metabolic pathways in the *numbers* of ATPs generated per glucose-unit input and these are summarized in Table 3.1. This information can also be given as an energy efficiency:

$$k_b = \frac{\text{ATP output (joules)}}{\text{Glucose input (joules)}} \times 100$$

and this quantity is often referred to as **biochemical efficiency**. The ranking of the biochemical efficiencies of the metabolic pathways is:

aerobic system > fumarate systems > glycolytic−lactate and opine
systems

Table 3.1. Efficiencies of ATP production by various metabolic pathways

System	No. ATP/ glucose unit input	k_b (%)	η^*
Glycolysis lactate opine	2	3	0.60
Fumarate	4-6	5-6	>0.70
Aerobic	36	37	0.66

Data largely from Gnaiger (1983, Table 2)

Not all the energy losses from the glucose input are as heat, some are waste products such as lactate and opines, and it is possible to calculate more precise thermodynamic efficiencies (symbolized as η^*) from the ratio of ATP output power to catabolic input power (Gnaiger, 1983; 1986). Again these are summarized in Table 3.1 and this time the ordering is:

fumarate systems > aerobic system > glycolytic–lactate and opine systems

If, as is often the case, waste products are lost from organisms as excreta, then it is likely that k_b is more relevant for selection than η^*, because the former takes into account total energy losses, not just energy losses as heat. *A priori*, it is not unreasonable to expect that selection will have tended to minimize energy losses and hence to maximize k_b, for then less energy is needed to generate the required ATP and more energy is available for other fitness-promoting metabolic processes such as production. The evolution of aerobic metabolism, as an exploitation of more efficient metabolism allowed by the appearance of oxygen on earth, is compatible with this expectation. But, why have the inefficient glycolytic–lactate and opine systems not been replaced by the more efficient fumarate systems for ATP production in the absence of oxygen? A plausible explanation is that the more efficient pathways are subject to a rate constraint; they generate ATP considerably more slowly than the less efficient processes (sometimes by a factor of more than 100-fold according to Livingstone, 1983) and this is probably based on a fundamental, thermodynamic trade-off between metabolic speed and efficiency (Gnaiger, 1983, 1986 and see Fig. 3.3

but cf. Watt in press). Since the rate of ATP production might be more important than efficiency, e.g. to maximize activity for escaping a predator or chasing a prey, and to maximize rate of production (see below), efficiency need not always be maximized.

Thus the glycolytic–lactate and opine pathways have evolved to meet needs for bursts of high activity which can cause particular muscles to run out of oxygen for short periods. However, this activity is not sustained and occurs only occasionally, so efficiency is relatively unimportant. On the other hand, in hypoxic and anoxic environments which involve the whole animal, not just some of its muscles, there is likely to be more selection pressure for efficiency — what Gnaiger (1983, 1986) calls e-selection as opposed to p-(power)-selection.

As already noted, aerobic pathways give highest ATP per total substrate input (highest k_b) but not necessarily per catabolic power input (η^*) (Table 3.1). Further costs for aerobiosis are attributable to maintaining those mechanisms, ranging from behavioural and morphological to molecular, that are concerned with the provision of oxygen. Interestingly, in vertebrate animals there is a distinction between so-called red muscle fibres that work continuously at a steady rate under aerobic conditions, and contain the red oxygen-storage pigment, myoglobin and so-called white fibres, which are specialized for anaerobic, burst work and do not contain myoglobin. This division of labour may

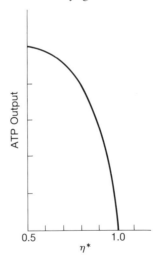

Fig. 3.3. The trade-off between ATP output power and η^* for a stoichiometrically fully-coupled system of ATP generation in steady state. The curve is deduced on thermodynamic grounds. Redrawn from Gnaiger (1983, Fig. 9); see Gnaiger (1986) for a more elaborate version.

minimize the costs of maintenance of the whole system. Presumably respiratory structures, such as gills and lungs, circulatory structures such as hearts, vessels and the blood itself, and ventilatory behaviours are optimized on the same basis — to minimize costs. Unfortunately, we do not have sufficiently precise information on the costs and benefits involved here to make good models but there have been some preliminary attempts (e.g. Calow, 1981).

3.3   Costs of production

Let us assume, at least initially, that efficiency k_b is constant and that the ATP generated from the appropriate metabolic processes can be used to pay costs of maintenance (C_m) and costs of production (C_p). C_m will be dealt with below, but assume for the time being that it is fixed for a given body size. C_p is used to pay the costs of getting, ingesting and digesting food, and the costs of synthesis (assembly of macromolecules and organization into tissues), so it is not unreasonable to assume that C_p is some function of the rate of tissue production, P; say

$$C_p = b\ P \qquad\qquad (3.1)$$

Since $P = A - (C_m + C_p)/k_b$ (from Fig. 3.1) we have

$$P = A - (C_m + b\ P)/k_b \qquad\qquad (3.2)$$

where A = energy absorbed after digestion. This suggests the linear relationship between P and A that is shown in Fig. 3.4a; i.e. if A is less than C_m/k_b, P is negative, but once C_m is covered P becomes positive. To relate P to ingested energy, I, remember that $A = k_d\ I$, where k_d = digestive efficiency and remember that k_d itself is related to I, in general reducing as throughput increases (Chapter 2). Hence, this suggests a curvilinear relationship between P and I as shown in Fig. 3.4b, and this kind of relationship has indeed been recorded (Fig. 3.4c) for ruminants by Blaxter (1980).

   The relationship between gross production efficiency $(P/I = k_1)$ and I that follows from Figs. 3.4b and c is illustrated in Fig. 3.4d. On to this are also plotted, as broken lines, contours of equal power output (production rate); where production rate $= I\ k_1$.

   For reasons already noted in Chapters 1 and 2, maximized rates of production are likely to be optimum; i.e. the point on the curve

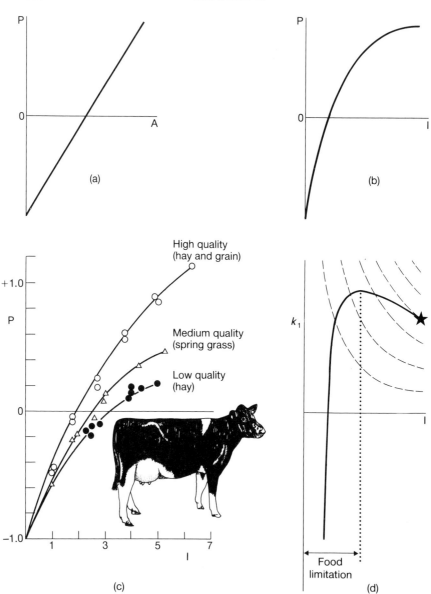

Fig. 3.4. Relationship between production P, energy absorbed from food A, and energy content of food I. Data in (c) are from Blaxter 1980. P and I were measured in multiples of fasting metabolism. In (d) broken lines represent contours of equal production. See text for discussion.

relating k_1 to I which makes contact with the highest power contour. The first thing to note from Fig. 3.4d is that, dependent upon the exact shape of the curve, this optimum need not be coincident with the maximum efficiency possible from a particular metabolic system. One analysis conducted by Ware (1975) on planktivorous fish, where both ingestion rate and metabolic output are directly proportional to swimming speed, indicates that fish swim at speeds that maximize growth rates, but not necessarily growth efficiencies or growth efficiencies per unit ingestion rate. The second thing to note is that food limitations may prevent the animal working at its physiological maximum.

Fig. 3.5 illustrates the possible effects of changes in metabolism. In transition i in Fig. 3.5a, C_m is fixed but C_p increases, so the efficiency per unit food ingested reduces. This could represent the shift to a more active feeding mode. It is favoured if the new feeding behaviour improves the rate of supply of food. For example, the lizards *Cnemidophorus tigris* and *Calisaurus draconoides* are ecologically related (therefore probably have the same C_m) but the former actively forages whereas the latter is more sedentary. In consequence *C. tigris* respires about twice as much as *C. draconoides* (i.e. due to elevated costs of supply, C_p) (Anderson & Karasov, 1981) and yet the former has a significantly greater foraging efficiency than the latter (Table 3.2).

There might also be behavioural and ontogenetic shifts between curves, if the returns (or likely returns) from one strategy are better

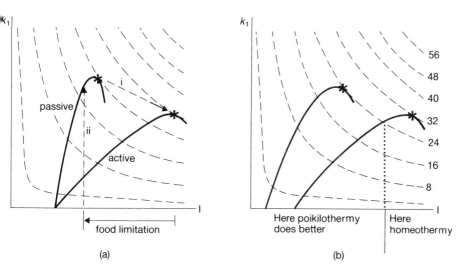

Fig. 3.5. Effects of changes in an animal's way of life (see text). Symbols as in Fig. 3.4d.

Table 3.2. Comparison of energy costs and gains of feeding in two lizards. Data from Anderson & Karasov (1981)

	(a) Feeding mode	(b) Average size (g fresh weight)	(c) Field metabolic rate J g^{-1}h^{-1}	(d) (c) as % resting laboratory metabolic rate	(e) Food ingested per feeding period (metabolizable kJ)	(f) Feeding period (hours)	(g) Foraging efficiency
Cnemidophorus tigris	Active forager	16g	22.5	150	3.7	5	2
Calisaurus draconoides	Sit and wait	9g	9	330	0.9	10	1.1

Foraging efficiency (g) $= \dfrac{\text{(e)}}{\text{(c)} \times \text{(b)} \times \text{(f)}}$

than those from another. An example of a behavioural change would
be a shift to a seek-out (active) strategy if the returns from a sit-and-
wait (passive) one are low relative to the expected returns of becoming
more active (transition i on Fig. 3.5a) or perhaps more likely a shift to
a sit-and-wait strategy if the food supply deteriorates (transition ii on
Fig. 3.5a). For example, Westerterp (1977) monitored activity and that
component of metabolic rate due to activity (C_a) simultaneously in
starving rats. The results are summarized in Fig. 3.6 and indicate not
only that activity and hence C_a reduced during food deprivation but
that C_a/(activity score) also reduces, suggesting that the rat engages in
less energetically expensive behaviour as it becomes more starved.

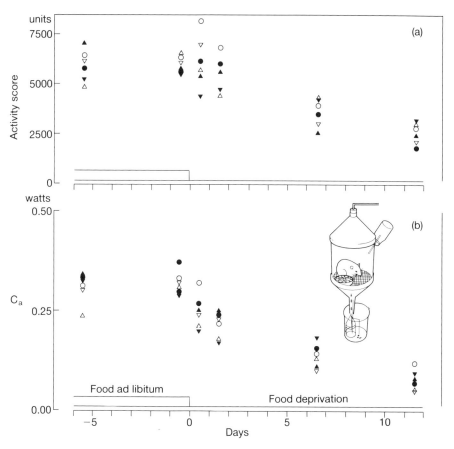

Fig. 3.6. Change of (a) daily activity score, and (b) C_a during food deprivation. Each
point represents one rat. From Westrterp (1977).

In Fig. 3.5b C_p is fixed but C_m increases. This could represent the evolution of homeothermy. Here a constant, above-ambient daily temperature is maintained by the metabolic generation of heat (endothermy), and is to be contrasted with the poikilotherm/ectotherm condition where internal body temperature fluctuates with external ambient temperature. There has been some controversy over whether homeothermy results from improved retention (by insulation) of endogenously generated heat or from the increased generation of heat (Calow, 1977a). However, the conversion efficiencies of homeotherms (k_1 and k_2) are generally lower than those of poikilotherms (as in Fig. 3.5b) suggesting that the latter is involved. For example, conversion efficiencies (k_1) of 25−35% were determined in young, growing homeotherms as long ago as the 1880s by Rubner and later by Kleiber (1933) and Mayer (1948). Since the average absorption efficiency is around 60−70% then k_2 ($=k_1/k_d$) is between 40 and 50%. The values of k_1 and k_2 for poikilothermic invertebrates are generally greater than these values (Table 3.3 and Wieser, 1985)

Because of the higher levels of metabolism and hence rates of food supply homeotherms can, as is suggested in Fig. 3.5b, maintain the same or higher levels of production than poikilotherms albeit at lower levels of efficiency. Thus, despite the fact that the conversion efficiencies of homeotherms are generally lower than poikilotherms (see above), their individual growth rates might be an order of magnitude greater than those of poikilotherms (Fig. 3.7). Population growth rate is also correlated positively with metabolic rate and is potentially greater in homeotherms than poikilotherms (Fenchel, 1974; McNab, 1980; Henneman, 1983).

Again, the shift between 'poikilotherm' and 'homeotherm' curves might be ontogenetic. For example young 'homeotherms' often have no temperature controls (i.e. are metabolically poikilothermic) deriving warmth from parents. The timing of the shift from poikilothermy to homeothermy could depend on food supply (Fig. 3.5b). For example, McClure & Randolph (1980) have shown that the small rodent *Sigmodon hispidus* becomes homeothermic at a smaller size and earlier age than a similar species *Neotoma floridana* and therefore expends more energy in respiratory metabolism (costs of living) and has a lower conversion efficiency k_2 but a higher growth (production) rate (Fig. 3.8). The metabolic strategy of *S. hispidus* has evolved under conditions of continuous food abundance or greatly fluctuating rations whereas that of *N. floridana* has evolved under chronic food shortage.

Table 3.3. Conversion efficiencies in selected invertebrates (after Calow, 1977b)

	k_1 (%)	k_2 (%)	k_d (%)	Units*
Homeothems (for comparison)	25−35	40−50	60−70	
Coelenterata				
Actinia equina	?	50−94	?	En
Hydra pseudoligactis	42	?	?	En
Platyhelminthes				
Dendrocoelum lacteum	?	*c.* 60−70	?	En
Mollusca				
Mytilus edulis larva	?	73	?	O
Mytilus edulis metamorphosed	?	84	?	O
Littorina littorea veliger	?	62	?	O
Nassa reticulata veliger	?	63	?	O
Arion rufus	32	55	65−80	En
Ostrea edulis larva	*c.* 51.1	78.6	50−80	En
Ancylus fluviatilis	45	75	60	En
Planorbis contortus	63	74	85	En
Annelida				
Tubifex tubifex	*c.* 32	64	*c.* 50	En
Nereis diversicolor	43	80	*c.* 50	En
Arthropoda				
Crustacea				
Artemia spp.	53	?	?	W
Calanus hyperboreus	50	89	*c.* 55	En
Calanus finmarchicus	36	68	60	D
Euphausia pacifica	27	30	90	C
Palaemon serratus larvae	?	76	?	En
Menippe mercenaria larvae	42	60	72	En
Rhincalanus nasutus	55	?	?	C
Calanus helgolandicus	72	?	?	C
Insecta				
Leaf-eating lepidopteran larvae	11−34	29−69	*c.* <50	D
Terrestrial predacious and blood-sucking insects	22−53	?	?	D
Blatella germanica	45	53	*c.* 85	En
Heriodiscus truquii	*c.* 36	*c.* 60	53−70	En
Lestes sponsa	35	80−90	*c.* 40−50	En
Pyrrhosoma nymphula	60	62	*c.* 90	En
Phytodecta pallidus	29	*c.* 60−69	*c.* 40−50	En
Arachnida				
Tarantula kochi	60	60	*c.* 100	D
Pardosa lugubris	42	42	*c.* 100	En
Stegnacurus magnus	66−76	74−96	40−70	En
Echinodermata				
Asterias rubens	95	?	?	En

* En = energy; O = organic material (ash-free dry weight); W = wet weight; D = dry weight; C = carbon.

55

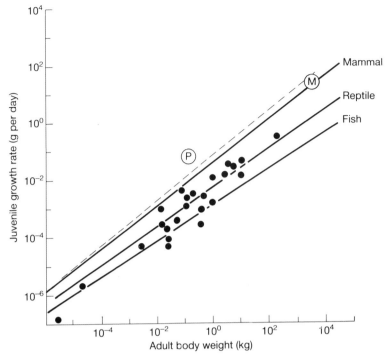

Fig. 3.7.   Growth rates of homeotherms compared with poikilotherms. Points are for invertebrates. P = endoparasites of homeotherms which can cash in on homeothermy of the host without paying the costs of endothermy. M is a cephalopod — both octopuses and squids have very high growth rates. The broken line represents the regression for the invertebrate data corrected to 38°C. After Calow and Townsend (1981b).

3.4   Costs of maintenance

Costs subsumed under the heading of maintenance are many and varied, ranging from ionic and osmotic control to the repair of damaged molecules and structures. It is, nevertheless, legitimate to consider these together because they are all essential for maintaining biomass in a viable state. Very fundamentally, to be viable, biomass has to be organized, and, in consequence, suffers thermodynamic injury. More specifically, biological organization needs power and involves the production of heat, which occasionally and accidentally damages the molecules which constitute tissue. There is, therefore, a continuous need for the repair and replacement of tissues and molecules and this manifests itself as the dynamic steady-state of the body (Schoenheimer, 1946) — molecules and cells turning over, even in non-growing adults.

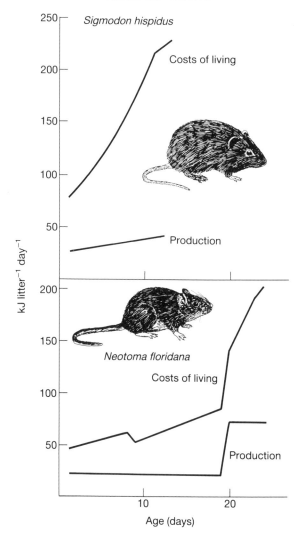

Fig. 3.8. Total chemical energy utilization (kJ·litter⁻¹·d⁻¹) for growth and respiration of young *Neotoma floridana* and *Sigmodon hispidus*. After McClure & Randolph (1980).

Over and above this, environmental stressors might increase the need for maintenance and hence maintenance costs. Thus increased maintenance costs have been documented in freshwater invertebrates living in fast-flowing waters (Fox & Simmonds, 1933), estuarine fishes suffering osmotic stress (Stearns, 1980) and invertebrates in some polluted environments (Bayne, 1976). In endotherms there is also a cost of maintaining body temperature. We have considered this in the last section and will not treat it further here.

The physiological processes involved in counteracting thermodynamic injuries and environmental stressors are active (e.g. transport against concentration gradients, cell division, protein synthesis) and require ATP. Repair processes can also involve replacement of damaged systems and hence require building blocks from the food resources. Hence, maintenance uses resources which might otherwise have been invested in production and there should be a trade-off between these two components of metabolism.

For example in the common mussel *Mytilus edulis* it is known that a particular gene (known as the Lap^{94} allele), which appears to be necessary in ordinary seawater, is wasteful in the excretion of nitrogen at lower salinities, causing reduced growth and increased mortality. As a result the Lap^{94} gene is outcompeted by other alleles at lower salinities, resulting in clines in allele frequencies which mirror clines in salinity. Such clines have been found in several estuaries in North America and in the Baltic (Bayne, 1986).

3.5   Trade-offs and their measurement

As already noted, the fitness benefits from the investment in production are in terms of developmental rates and fecundity (n). Developmental rates affect time to first breeding (t_j) and time between breeding attempts (t_a). Hence the benefits from production are in terms of t_j, t_a and n. The fitness benefits from investment in maintenance are in terms of juvenile survivorship, S_j, and adult survivorship between breeding attempts, S_a. Hence the allocation of resources between maintenance and production involves the following trade-offs:

1 S_j and t_j
2a S_a and t_a
2b S_a and n

Trade-off 1 implies that growth rates will always be submaximum. If there were no fitness costs involved, selection would maximize growth rates within any physiological constraint because this would minimize development time. Yet several lines of evidence suggest growth rates are not always at maximum levels:

1   After periods of growth retardation or inhibtion, growth rates often increase above 'normal' for the age/size class, suggesting that 'normal' is submaximum. These so-called catch-up phenomena (Prader *et al.*, 1963) will be considered further in Chapter 5.

2 Endocrine treatments can accelerate growth rates and, indeed, the very existence of endocrine controls suggests that growth rates are actually constrained below their physiological capacity by control systems that have evolved for this purpose.

Thus organisms do not adopt the growth pattern which is optimal according to the 'zero costs' hypothesis. This suggests that there are costs in growing faster and in the juvenile phase these can only be mortality costs, as in trade-off 1. These mortality costs might arise because: (a) growing faster might involve taking more risks to obtain more food; (b) it is conceivable that the chances of making mistakes in the developmental processes might increase as the rate of development increases; (c) faster growth implies that less is being invested in maintenance and repair. These possibilities are not mutually exclusive. On an interspecific basis, there is certainly a correlation between growth and mortality rates in echinoderms (Fig. 3.9), and a similar relationship occurs in fishes (Beverton & Holt, 1959). Organisms that continue to grow after breeding is initiated may also pay costs in terms of lost reproduction and again this will be considered in Chapter 5.

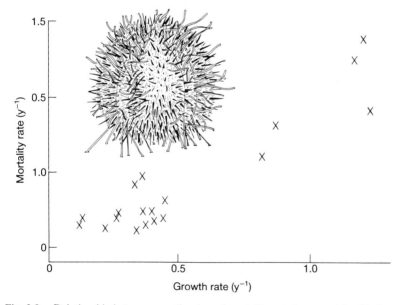

Fig. 3.9. Relationship between growth rate and mortality rate in sea urchins. Each point represents a different species. Growth rate was measured by the time constant in a growth equation. Redrawn from Ebert (1985).

There is no information available on trade-off 2a and for the time being we will have to assume that it exists. However, there is a considerable amount of circumstantial evidence for negative correlations between S_a and n predicted in trade-off 2b, at both an inter- and intraspecific level (e.g. see reviews in Calow 1979; Clutton-Brock, 1984; Reznick, 1985). One of the most compelling examples of this kind is provided by Clutton-Brock *et al.* (1983) who found increased mortality in red deer hinds that had produced a calf and suckled it compared with those which had not (Fig. 3.10). However, phenotypic correlations of this kind do not establish beyond doubt a causal relationship as implied by a trade-off (cf. the Principle of Compensation, Chapter 1; Sutherland *et al.*, 1986; Reznick, 1985; Bell, 1984a & b). Two other important lines of evidence are therefore from (a) physiological manipulation and (b) analysis of genetic correlations and selection experiments.

(a)   PHYSIOLOGICAL MANIPULATION

If there is a negative causal relationship between investment in reproduction and subsequent adult survival it should be possible to experimentally shorten the lives of long-lived, repeated breeders (iteroparous) by stimulating reproduction and to lengthen the lives of short-lived, single breeders (semelparous animals) by inhibiting it.

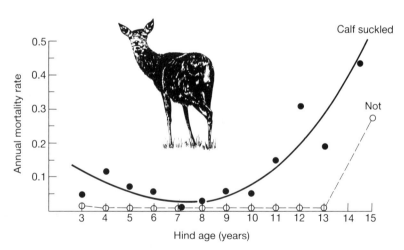

Fig. 3.10. Mortality cost of producing a calf in red deer hinds (after Clutton-Brock *et al.* 1982). See text for details.

The reproductive investment of male fruit flies has been manipulated in an elegant series of experiments by Partridge and Farquhar (1981), who provided males with either 8, 1 or no virgin females per day. Males provided with more virgin females increased their sexual activity accordingly, and this led to a progressive loss of lifespan (Fig. 3.11) after allowing for the effect of male size. Careful control experiments were carried out in case the number of companions affected male longevity even if the companions were sexually inactive, but because no such effects were found the control data (•) are not distinguished in Fig. 3.11. It is not easy to cause an above-normal reproductive investment in females, except possibly by manipulating clutch size (Nur, 1984a, see Fig. 4.4) or by hormone treatment, and even then the effects of the hormones may be so non-specific that the exact causal basis of the consequences might be difficult to identify. On the other hand, stopping reproduction is relatively easy; e.g. by preventing mating and by gonadectomy. These techniques have been used on a variety of organisms and they invariably have the effect of causing an increase in life-span. The data are summarized in Table 3.4.

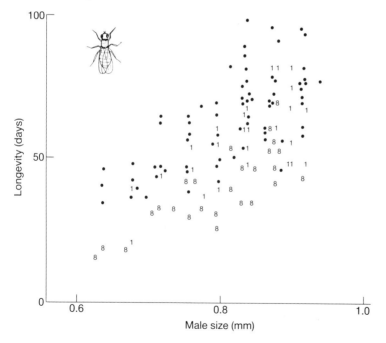

Fig. 3.11. Male longevity increases with size (measured as thorax length) in fruitflies. Longevity was decreased if males were provided with 1 virgin female per day, and further decreased if provided with 8 virgin females per day compared with controls (•).

Table 3.4. Evidence showing that the prevention of reproduction significantly increases the longevity of semelparous females (from Calow, 1977a)

1	Minnows normally die after spawning, but non-spawners may live an extra year.
2	Gonadectomized salmon do not spawn and are able to return to the sea.
3	Heat sterilized, unmated and ovariless *Drosophila* live significantly longer than mated females.
4	Virgin female bugs (Hemiptera) and beetles (Coleoptera) live longer than mated females.
5	Observations 3 and 4 are typical for most insects.
6	Death in some annual plants (e.g. tomatoes) can be postponed indefinitely if reproduction is prevented.

(b)   GENETIC CORRELATIONS AND SELECTION EXPERIMENTS

For a particular organism (or, more precisely, genotype) a high invest-ment in reproduction should lead to a reduced survivorship, but be-tween organisms those capable of higher fecundity might be capable of higher survival, e.g. because they lived in better environments. Hence, if we plotted survival against fecundity in a random sample of indivi-duals from a population we might obtain a positive relationship, though for particular genotypes there were in fact negative relationships (Fig. 3.12). It is, therefore, necessary to distinguish the within- from the between-genotypes effect, and this is possible only by calculating cor-relations of traits on individuals separated into groups of relations (parent−offspring, full sibs, half sibs etc.). These are known as **genetic correlations**. This kind of analysis also gives valuable information on the relative importance of environmental and genetic effects on quanti-tative traits; for example, genotype A in Fig. 3.12 might have been in a better environment than B and the good conditions might have had beneficial effects on both traits (for details of methods see Falconer, 1981).

The most thorough study so far reported on genetic correlations between S_a and n is that of Rose and Charlesworth (reviewed in Rose, 1983; 1984) on a laboratory population of *Drosophila melanogaster*. The general picture they obtained was of mainly negative genetic correlations. On the other hand phenotypic correlations were generally positive, possibly for reasons similar to those stated at the end of the last paragraph.

Clearly if there are negative genetic correlations, then artificial selection for an increase in one of the traits should cause a reduction in

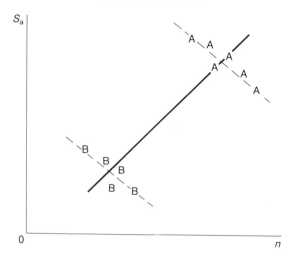

Fig. 3.12. Trade-off between fecundity (n) and survivorship S_a, illustrating genetic and environmental effects. Genotypes A are in a better environment than genotypes B, so have higher values of n and S_a, so the pooled data show a positive correlation between fecundity and survivorship. However genetic analysis of either population shows a negative relationship. One way to carry out a genetic analysis is to plot n (measured in mothers) against S_a (measured in daughters). If a relationship is found, it must be due to gene effects, since environmental effects on mothers and daughters are randomly different (but see Nordwijk, 1984). Falconer (1981) gives further details.

the other and vice versa. Again, some of the best work on this is the analysis by Rose and Charlesworth of *D. melanogaster* (Rose, 1983; 1984, — independently confirmed by Luckinbill *et al.*, 1984, see Fig. 3.13). Here, as expected, selection for late fecundity caused a significant lengthening of life, at the expense of early fecundity. Selection for earlier age of reproduction has also been carried out by Sokal (1970) on the flour beetle, *Tribolium castaneum*, and there too it had the effect of accelerating senescence.

These genetic correlations may be due to pleiotropies (i.e. genes affecting more than one trait) or other genetic mechanisms, and we here suppose that negative genetic correlations reflect physiological trade-offs. Of course this is only partially correct; because genetic correlations also depend on phenomena such as linkage which depend in turn on the organization of the chromosome. However, the trade-offs are based on fundamental physical constraints associated with resource allocation and cannot be modified further by natural selection. Other components of genetic correlations, on the other hand, can in principle be subject to modifications by selection and in this sense can be considered less fundamental. It is perhaps worth reiterating, at this

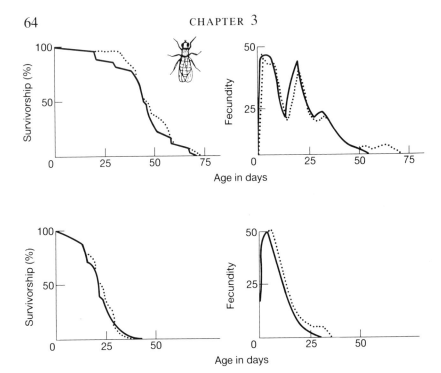

Fig. 3.13. Survivorship and fecundity schedules for *Drosophila melanogaster* after 13 generations of selection for late reproduction (top row) or early reproduction (bottom row). Dotted line indicates a replicate. Selection for early fecundity achieved a slight increase in early fecundity (bottom right) at a considerable cost in terms of survivorship (bottom left). Flies were maintained in milk bottles at constant adult densities of 50 pairs per bottle. After Luckinbill *et al.*, 1984.

point, our tacit view that the physiological interactions that form the subject-matter of this book are fundamental in developing a **general theory of organismic evolution**; information on genetic correlations is *necessary but not sufficient* for the testing of this theory.

OPTIMUM ALLOCATION IN MAINTENANCE

Given the trade-offs 1, 2a and 2b an optimum balance has to be struck between the physiological processes behind maintenance and production and this will be influenced by other environmental conditions. The optimization analyses will be carried out in Chapters 4 (for 2a and 2b) and 5 (for 1). There it will also be shown that the precise outcomes depend upon (a) exactly when costs are paid (for 2a and 2b) and (b) on the detailed form of the trade-off — e.g. we need to know far more about S_a and n than that they are negatively correlated; we need to know the shape of the trade-off curve.

3.6   Conclusions and summary

The statement that costs of living should be minimized and production maximized is seductive in its simplicity but ignores too much, for the two are not necessarily exclusive and costs of living can bring important survival gains. Evolution has apparently proceeded towards metabolic pathways that maximize the amount of ATP produced per substrate used but not necessarily to the minimization of heat production per ATP produced. However, because efficiency and rate of ATP production might be involved in a trade-off there are times when it is better to sacrifice efficiency, for example to power short-term bursts of activity which only occur occasionally. An understanding of production costs can give insight into the relationship between production rate and ingestion rate, and into the evolution both of different foraging patterns and of homeothermy. Costs of maintenance cause reduced growth and/or reproductive rates, but can bring substantial survival gains.

Methods of demonstrating costs and identifying trade-offs are discussed in some detail, and some pitfalls are pointed out. Despite being based on fundamental physical and chemical constraints these trade-offs have proved surprisingly difficult to expose. Experimentally, one usually hopes to map them by changing one variable and observing the response of others and, in principle, this may be achieved by physiological (e.g. hormone) treatment, clutch size manipulation, breeding experiments and genetical selection. A necessary assumption, though, is that the organism's responses will lie on the trade-off curve and this need not be the case because not all relevant genes may be present in the population. Thus a curve may be discovered which represents some evolutionary options, but not the whole range of options that would become available in a natural population on an evolutionary timescale. It is sometimes useful to call the set of evolutionary options the **options set** (example in Fig. 2.2), the boundary of which is the trade-off curve. Some understanding for what might be expected on the basis of *a priori* physiological insight can be helpful here (p. 58). Hence, to identify at least some of the evolutionary options open to organisms it is necessary to use a combination of carefully-designed experiments tempered with good theoretical background. Optimization models of the trade-offs identified in this chapter will be addressed in future chapters.

4

Reproduction

4.1   Introduction

More obviously than at any other part of the life cycle, the allocation of resources during reproduction influences fecundity (n), survival (S) and life-cycle timing (t). S and n are likely to increase but t to reduce as more resources are invested in the processes that affect them. Yet, as noted in Chapter 1, the resources available for allocation are limited and this is likely to lead to trade-offs between the various terms.

4.2   Two types of costing

Imagine an organism which is born at time $t = 0$, and develops for a period t_j before producing and releasing n propagules. There is then a period, t_a, before a further bout of breeding and so on (Fig. 4.1a).

The chances of the organism surviving from birth to breeding are S_j, and between breeding seasons, S_a. In this model it is assumed that because of the trade-offs referred to in Chapters 1 and 3, investments in n cause reductions in S_a and lengthening of t_a.

Now imagine a slightly more complex model where there is a period, *before* the release of propagules, when the organism is preparing for reproduction. Call this t_a, and for convenience let this equal the time between subsequent breeding bouts. (This assumption could be relaxed without affecting our conclusions.) Now on the basis of trade-offs similar to those referred to above it can be appreciated that an organism might pay the costs of reproduction prior to the release of propagules (Fig. 4.1b).

By analogy with accounting, the first model is referred to as **absorption costing**, since the cost of the product (propagules) is inventoried and only realized on the 'sale of the goods' (release of the propagules), and the second model is referred to as **direct costing** when (at least part of) the cost is not inventoried but set against revenue immediately (Sibly & Calow, 1984). Note that the important point in the life cycle from the point of view of this classification is when propagules are *released* and become independent of parents. This point might be the release of gametes or of brooded offspring.

Costs of reproduction might be incurred *before* gamete release as a result of: (a) increased risks associated with courtship and copulation due either to (i) increased conspicuousness to predators or (ii) aggressive behaviour to sexual competitors (e.g. Tinkle, 1969); (b) increased foraging effort needed to obtain resources to make the gametes and/or to prepare nests etc.; (c) morphological distensions and mechanical imbalances caused by the accumulation of gametes or their precursors or carrying the young, again causing increased metabolic requirements and mortality risks (e.g. if pregnant females are more easily caught by predators); (d) the drain of resources from somatic to gametic tissues and processes (reviewed in Calow, 1979). Of all these, (d) is least likely as a direct cost since, at least in principle, it is reversible and there is considerable evidence for gamete and embryo resorption by parents under stress (e.g. reviewed in Calow, 1973).

Costs might be incurred *after* gamete release as a result of: (a) the risks of parturition either (i) because individuals become more conspicuous or vulnerable to predation and disease at these times or (ii) due to direct injury resulting from the release of gametes or young (the

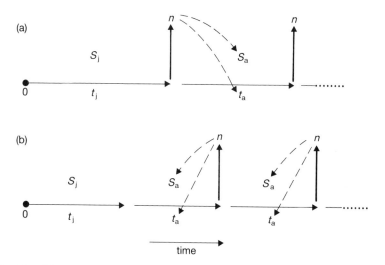

Fig. 4.1. Diagrams showing two types of life cycle considered in Chapter 4. Broken lines represent influences. (a) absorption costing, in which the costs of producing n offspring are paid after reproduction; (b) direct costing, in which costs are paid before reproduction. Symbols are defined in the text.

most extreme example of this is in some insect species where progeny eat their way out of their mothers' abdomens); (b) the loss from the parent of resources locked into the reproductive products which are then no longer available as a fail-safe even under stress (cf. above); (c) the accumulation of the effects of metabolic stress arising from the events leading up to the release of gametes. Thus at any one time a migrating adult salmon may not be over-stressed with regard to its own metabolic capabilities and yet the cumulative effect of the whole migration in association with the release of eggs is devastating; (d) care of eggs and offspring, after they are released involves considerable metabolic costs (Bryant, 1979) and survival risks (Nur, 1984 a & b, see p. 74; Hussell, 1972; Askenmo, 1979) when it occurs.

Just as the distinction between absorption and direct costing has an important effect on the statement of profitability in accountancy so the different ways of accounting the costs of reproduction make a difference to predicted life-cycle optima. We have already had a little to say about this in Chapter 1 (p. 22). In the analyses that follow we first concentrate on absorption costing and then, in a final section, indicate how these conclusions are modified in the case of direct costing. There are four reasons for this emphasis: (1) the mathematics of absorption costing are a little more straightforward than those of direct costing; (2) they are also more familiar, since most of the literature on lifecycle theory is in terms of absorption costing; (3) though all organisms have to prepare for reproduction so that some level of direct costing always applies, it seems probable that there will have been selection against direct costs having a dominant effect because, as breeding is such an important component of fitness, it will often be advantageous to put off the full cost of reproduction until after propagules are released. Clearly, excessive direct costs could lead to no viable offspring being released (hence minimal fitness). It is hard to imagine therefore how any semelparous organism could indulge in direct costing; (4) it will be clear from the above description of direct and absorption costs that the former are largely behavioural and the latter physiological. Throughout this book we emphasize the physiology and only point, where appropriate, to possible behavioural complications.

4.3   Possible trade-offs and their causes

We can define F, the usual fitness measure (p. 10), in terms of the five life-cycle variables defined above most simply by assuming that there are two classes of individual, juvenile (j) and adult (a), and that life-cycle variables differ between but not within classes. Then, (Box 4.1), the Euler/Lotka equation (p. 10) can be divided into two age-specific components:

$$\text{for absorption costing: } 1 = \tfrac{1}{2}e^{-Ft_j} S_j n + e^{-Ft_a} S_a \qquad (4.1)$$

Here the first term contains the juvenile variables (time to first breeding (t_j), survival over this time (S_j)) and fecundity (n); and the second term contains the adult variables, describing time between breeding (t_a), and survival over this period (S_a). The point of the equation is to show how these variables determine fitness (F), which is (unfortunately!) present in both terms. S can be replaced where necessary by a mortality

---

BOX 4.1. Derivation of equation 4.1.

For the life cycle shown schematically in Fig. 4.1a, the probability of animals surviving to age t_j, when they produce n offspring, is S_j; the probability of surviving to age $t_j + t_a$, when they produce another n offspring, is $S_j S_a$; the probability of surviving to age $t_j + 2t_a$ is $S_j S_a^2$, and so on. Substituting these values into equation (1.3) gives

$$1 = \tfrac{1}{2}e^{-Ft_j}S_j n + \tfrac{1}{2}e^{-F(t_j+t_a)} S_j S_a n + \tfrac{1}{2}e^{-F(t_j+2t_a)} S_j S_a^2 n + \ldots$$

Setting $e^{-Ft_a} S_a = y$ this can be rewritten

$$1 = \tfrac{1}{2}e^{-Ft_j} S_j n (1 + y + y^2 + y^3 + \ldots)$$

$$= \frac{\tfrac{1}{2}e^{-Ft_j} S_j n}{1 - y},$$

using the formula for the sum of a geometric series (Causton, 1983). Hence

$$1 = \tfrac{1}{2}e^{-Ft_j} S_j n + y,$$

$$\text{i.e. } 1 = \tfrac{1}{2}e^{-Ft_j} S_j n + e^{-Ft_a} S_a. \qquad (4.1)$$

rate (μ) if we make the assumption that deaths occur according to a Posisson process (i.e. randomly) such that $S = e^{-\mu t}$. Then

$$1 = \tfrac{1}{2}e^{-(F+\mu_j)t_j} n + e^{-(F+\mu_a)t_a} \tag{4.2}$$

We refer to the Ss, μs, ts and ns as life-cycle variables.

In keeping with the previous philosophy, we expect natural selection to favour allocation strategies that lead to combinations of S (or μ), n and t that maximize F (in the local sense defined previously). If there are no constraints, then it is intuitively clear that fitness is maximized by maximizing n, S_a and S_j and minimizing t_a and t_j (Sibly & Calow, 1983). However, as already stated, (p. 23) trade-offs are inevitable and optimum combinations of life-cycle variables will depend upon the nature of these. It is easiest to take them into account by considering variables two at a time.

Using the landscape approach introduced in Chapter 1, the 5 life-cycle variables S_a, S_j (or μ_a, μ_j), t_a, t_j and n can be considered in 10 possible pair-wise combinations and these are defined in Table 4.1. Some of these variables refer specifically to the growth phase of the life cycle and others to the interaction between growth and reproduction, once the latter has been initiated. These will be discussed in the next chapter and are not considered further here. All the remaining trade-offs can be explained in terms of the utilization of limited resources (i.e. on the basis of physiology) and will be treated in this way in the

Table 4.1. Pair-wise combinations of S_a, S_j, t_a, t_j and n

	Trade-off	Interpretation	Chapter	p.
1	n vs. S_a	Risky reproduction	4	71
2	n vs. S_j	Offspring numbers or offspring quality	4	77
3	S_a vs. S_j	Parental care	4	81
4	S_a vs. t_j	Parental care	4	81
5	S_j vs. t_j	Growing fast is risky	5	92
6	S_a vs. t_a	Shortening the interbreeding period is risky	*	
7	n vs. t_j	More smaller vs. fewer larger offspring	4	84
8	n vs. t_a	Adults with more offspring take longer to recover	*	
9	t_a vs. t_j	Parental care	4	81
10	t_a vs. S_j	Parental care	4	81

* not considered

sections that follow. However again it is important to appreciate that some of the trade-offs might also have a behavioural cause. For example, n depends on finding a mate and a suitable site for egg-laying, S_a and S_j on the ability of organisms to stay out of danger, and t_a and t_j depend upon feeding strategies. Hence, the trade-offs may derive from physiological and behavioural factors acting either jointly or in isolation. In what follows we shall, as before, concentrate on physiological factors.

4.4 Trade-off between investment in reproduction and post-reproductive survival of parents (n versus S_a)

Equation 4.1 can be simplified further by assuming that $t_a = t_j = 1$. This would be reasonable, for example, for organisms that breed at annual intervals from age one such as many invertebrates, fishes and birds. Then

$$e^F = \tfrac{1}{2} S_j n + S_a \tag{4.3}$$

so

$$F = \log_e (\tfrac{1}{2} S_j n + S_a) \tag{4.4}$$

From this we can construct a selective landscape in terms of n and S_a (Fig. 4.2a). It is quite clear that, other things being equal, fitness (F) should increase with S_a and n, and the equations show this will occur at a reducing rate. Moreover, provided $\tfrac{1}{2} S_j n + S_a$ is constant, F will remain constant, so different combinations of S_a, S_j and n can give the same fitness. Focus on S_a and n, with S_j held constant. To keep a constant F, an increase in n can be compensated for by a proportional (linear) reduction in S_a and vice versa. The extent of the required adjustment (i.e. proportionately) depends upon juvenile survival, for with bigger values of S_j more adjustment is required in S_a to balance changes in n. Hence a two-dimensional projection of the surface in Fig. 4.2a on to the $n-S_a$ plane yields a map of the surface with contours of equal F represented as straight lines with a negative slope of $S_j/2$ (Fig. 4.2c). As juvenile survivorship increases, the slope of the contours also increases. (Fig. 4.2b & d).

In principle, selection should favour allocation patterns that maximize F, i.e. are out on the furthest contours to the right in Fig. 4.2c and d. However, it has already been noted that trade-offs are likely between S_a and n (Chapter 3). These represent the outer boundaries of

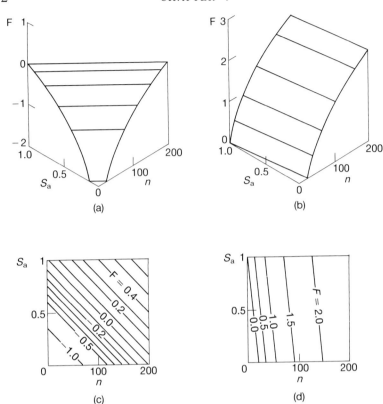

Fig. 4.2. (a) and (b) selective landscapes, i.e. fitness, F, plotted as a function of n and S_a. (a) shows the landscape for low offspring survivorship, $S_j = 0.01$, and (b) for high S_j = 0.1. Contour maps of (a) and (b) are shown respectively in (c) and (d).

sets of combinations of S_a and n that are physiologically feasible. Hence the optimum combinations of feasible combinations of S_a and n must occur on these boundaries, which we call **trade-off curves**, so it can be appreciated that the form of these curves is important in determining the optimum allocation strategy. Several theoretical trade-off curves are shown in Fig. 4.3. Optima are points on these curves which make contact with the highest contours of F (for a more formal treatment of this in terms of μ rather than S, see Box 4.2). The linear trade-off (a) and the curvilinear trade-off (b) predict no reproduction at all or maximum investment depending on the slopes of the contours. Curvilinear trade-off (c) predicts either a little or a large investment in reproduction, again depending on contours (and will be considered in more detail below). Finally, curvilinear trade-offs d−f allow, at least in

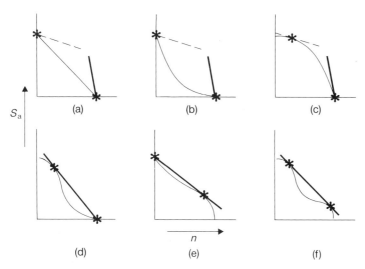

Fig. 4.3. Possible forms of the n vs. S_a trade-off. Straight lines are fitness contours, solid and broken lines correspond to high and low juvenile survivorship respectively * are optima. See text for discussion.

principle, for there to be more than one optimum for a single value of F.

The first point to appreciate is that since natural selection will, by definition, have restricted the expression of the trade-off curve to an optimal point or sector, then not much is likely to be learnt about its overall shape from straightforward correlations, within- or between-populations, between S_a and n (as discussed in Chapter 3). Experimental manipulations, taking organisms outside the normal expression of their trade-off curves, should be helpful, and one of the few extant examples is provided by Nur (1984 a & b).

In 1978 he manipulated the clutch sizes of 106 broods of blue tits (*Parus caeruleus*) breeding in Wytham Wood near Oxford, UK, to experimentally create broods of sizes 3, 6, 9, 12, and 15. The following winter, mist-nets were placed throughout the wood to catch adults, and breeding adults were also caught in nest boxes the following spring. The number of birds recaptured was taken as an index of S_a, and this is plotted for females in Figure 4.4. Note that the trade-off curve appears to be concave seen from above, and remember from Fig. 4.3b that the optimal strategy is then to invest maximally in one suicidal act of reproduction. However blue tits breed repeatedly over several seasons. Nur suggests this may be because two assumptions of the theory have

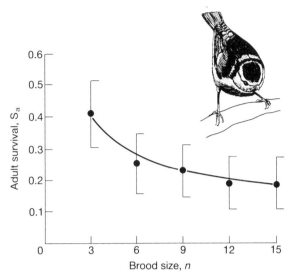

Fig. 4.4. Survival of adult females in the year after breeding plotted against experimentally-manipulated brood size. After Nur (1984a).

not been met. Firstly, the cost of producing eggs has been ignored, and secondly it is not taken into account that if a parent dies within 5 weeks of hatching the chances of offspring surviving are severely reduced or nil. It should also be noted that there may be variation in the form of the trade-off curve between years and that males might have different trade-off curves from females.

Some idea of the likely forms of trade-off can also be obtained *a priori* from an understanding of resource allocation. The simplest possibility is if resources are allocated to reproduction by parents on the basis of somatic physiological priorities; the first few gametes being formed from resources in excess of requirements (e.g. stored specifically for reproduction) and entailing no parental mortality risks (mortality rate = 0); the next from resources needed by but not essential to metabolism (mortality risks increase; $\mu \rightarrow +$); the next from resources essential to metabolism (death becomes a certainty; $\mu \rightarrow \infty$). This is illustrated in Fig. 4.5a and the consequent trade-off of S_a and n in Fig. 4.5c. In the latter, S_a reduces at an increasing rate with n. Another slightly different model, advanced by Ward *et al.* (1983) suggests that after use of the storage materials, μ rises continuously with increasing n, but *always remains finite*; i.e. there is no stage at which death becomes a certainty. The resulting relationships between μ_a & n and S_a

& n are illustrated in Figs. 4.5b and d respectively. Note the similarity between Fig. 4.5d and Fig. 4.4.

A possible way of distinguishing between these alternatives is to assume that resources are used in gamete formation according to the same set of physiologial priorities which operate under conditions of starvation (Calow, 1984). If this is true, survivorship curves for cohorts of starving animals should give some clue as to the form of the trade-off curve. Fig. 4.6a, shows data on this for the fry of 12 species (taken from Ivlev, 1961). In some species there is an initial period of high mortality, but this probably represents the death of 'weaklings'. If this is ignored the curves are rather ambiguous. In some, survival reduces continuously to zero whereas in others there is a some 'slowing-off' as starvation continues so that curves are pulled to the right. However, visual inspection suggests that this is never very marked. Lorraine Maltby has carried out similar experiments on adult *Asellus aquaticus* (freshwater isopods) but again results to date are somewhat ambiguous. In one population results conformed to the model in Fig. 4.5c (Fig. 4.6b) and in another, to the alternative sigmoid model in Fig.

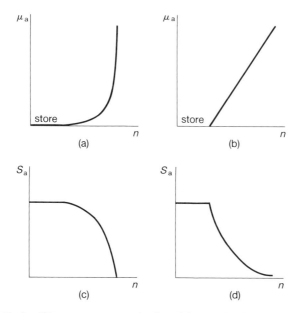

Fig. 4.5. Trade-off between post-reproductive adult mortality (μ_a) (and survival (S_a)) and fecundity (n). (a) (c), and (b) (d) are different cases. (a) (c) is from Calow & Sibly (1983) and (b) (d) from Ward *et al.*, (1983).

4.5d (Fig. 4.6c). Clearly more work is needed on survival under starvation, particularly on older animals.

Optimum combinations of S_a and n, i.e. life cycles, are nevertheless, most easily defined for the curve in Fig. 4.3c for here there is only one point that will be tangential to a linear contour so the solution is unique. In this case it is clear that the optimum will depend upon the slope of the contour and hence on S_j (see above). With a shallow slope (juvenile survival poor) low n, with good chances of post-reproductive survival, and hence repeated breeding (**iteroparity**) is optimum (see broken line). With a steep slope (juvenile survival good) high n, with poor chances of post-reproductive survival (i.e. a single breeding episode — **semelparity**) is favoured (see solid line).

Adult mortality factors such as accident, disease and predation over and above those due to reproduction, are also likely to be important in the selection of reproductive patterns. Calling the former **extrinsic factors** (survival chances as a result of these $= S_{ex}$) and the latter **intrinsic factors** (survival chances as a result of these $= S_{in}$) then the overall survival chances (S_a) are obtained from the product of the two operating in isolation (probability of 2 events occurring simultaneously):

$$S_a = S_{in} \, S_{ex} \qquad (4.5)$$

with lower S_{ex} the trade-off curve between S_a and n moves 'down' but

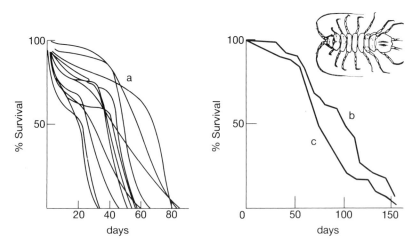

Fig. 4.6. (a) Survival (%) of fry of 12 species of fish. From Ivlev (1961). (b) and (c) are for two populations of *Asellus aquaticus* and were supplied by L. Maltby.

the extent of the shift is greater for higher values of S_{in}. Hence the two curves converge to zero (Fig. 4.7). For the same juvenile mortality (S_j) adults that suffer higher extrinsic mortality should invest more in reproduction. Note that for reasons already discussed in Chapter 1, there may be complications from the Principle of Compensation affecting this aspect of the model. However, if it operates here it will tend to enhance the effects of changing S_a. These predictions on the basis of S_j and S_{ex} are tested against information from field studies in Chapter 7.

Finally, we should re-emphasize that all the predictions made above depend crucially on the form of the trade-off curve. We have emphasized curve Fig 4.3c since it is a very plausible one, but others are also plausible and may be at work in particular populations, especially if behavioural factors dominate (Calow, 1985). Another major complication is stochasticity in the form of the trade-off and in the position of fitness contours, and we shall return to this later.

4.5 Trade-off between fecundity and survival of offspring (n vs. S_j)

The trade-off between n and S_j has been widely discussed from the point of view of predicting optimal clutch size in birds (Lack, 1954). Where offspring survival depends on feeding by parents, as in altricial birds, individual offspring survivorship generally declines if the number of offspring is increased because with more offspring there is a greater chance of a shortfall of food, and therefore of starvation. S_j is then a decreasing function of n.

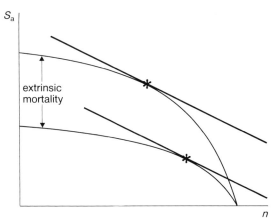

Fig. 4.7. Extrinsic adult mortality lowers the $S_a(n)$ trade-off curve and shifts the optimum reproductive investment (starred) to the right.

BOX 4.2. Trade-off between n and μ_a.

The trade-off between n and S_a considered in Fig. 4.3 is partly due to a trade-off between n and μ_a and partly to a trade-off between n and t_a. This distinction becomes important if t_a can vary. If μ_a depends on n (e.g. because greater mortalities are endured by more fecund genotypes) then μ_a is a mathematical function of n which can be written $\mu_a(n)$ and represented as a curve as in Fig. 4.5. Using the method of Box 1.3 we know that the optimal strategy is characterized by the slopes of the trade-off curve $\mu_a(n)$ and the fitness contours, these being equal. Fitness contours are defined by equation 4.2 when only μ_a and n are allowed to vary. Differentiating equation 4.2 with respect to n,

and setting $\dfrac{\partial F}{\partial n} = 0$ to find the optimal strategy, we get

$$0 = \tfrac{1}{2}e^{-(F+\mu_j)t_j} - e^{-(F+\mu_a)t_a}\frac{\partial \mu_a}{\partial n}t_a$$

$$\text{i.e.} \quad \frac{\partial \mu_a}{\partial n} = \frac{S_j}{2\,S_a t_a}e^{-Ft_j+Ft_a} \qquad (4.6)$$

since $S = e^{-\mu t}$. Equation 4.6 gives the slope of the trade-off curve characteristic of the optimal strategy.

From equation 4.1 it is clear that if all other terms are ignored, $S_j n$ defines a fitness contour in the n vs. S_j selective landscape (since if it is constant, F is constant). In addition, F will increase with $S_j n$ (this can be proved rigorously; Sibly & Calow 1983). The Principle of Compensation can be ignored because compensation could occur through adjustment of either μ_a or t_a without any effect on optimal strategy. Hence, fitness is maximized by maximizing $S_j n$; i.e. *choosing the clutch size that results in the most offspring surviving to maturity*. This was first stated by Lack (1954) and hence is often referred to as **Lack's Hypothesis**.

Klomp (1970), in an exhaustive review, concluded that data sufficiently detailed and comprehensive to evaluate the hypothesis were

Table 4.2 Optimal clutch size in the swift *Apus apus* at Oxford UK

Year	n	No. of broods	No. of young dying	Per cent lost	$n S_j$
1958	2	21	2	5	1.9
	3	4	1	8	2.8
	4	2	4	50	2.0
1959	2	15	0	0	2.0
	3	4	0	0	3.0
	4	4	5	31	2.8
1960	2	18	2	6	1.9
	3	6	4	22	2.3
	4	5	14	70	1.2
1961	2	18	1	3	1.9
	3	6	4	22	2.3
	4	5	13	65	1.4

Data from Perrins (1964) omitting broods of 1 which were almost always successful. The broods of four were not natural but due to the experimental addition of nestlings from other broods. The normal clutch size of the swift at Oxford is 2 or 3 eggs. Lack (1966) considered the main evolutionary reason for this is that 3 is advantageous in fine summers when food is abundant, but disadvantageous (lower $n S_j$) in poor summers. See Nur (1984b) for a thorough evaluation of the optimality of clutch size

available for 7 species of birds. In six of these (great tit, swift, Laysan albatross, Manx shearwater, Leach's petrel and red-footed booby) the most productive brood-size was at or near the most frequent brood-size. An example is given in Table 4.2. In the remaining case (the gannet) the most frequent brood-size was lower than the most productive one. This might be explicable on adaptive grounds if, for example, rearing an enlarged brood is costly in terms of adult survivorship, so that it is only worthwhile for artificially-enlarged broods (see p. 71). However, Lack (1966) and Nelson (1964) thought the most probable explanation was that feeding was easier for gannets at the time of the experiments than it had been in the past.

Lack (1954) realized that the model could be extended from the case of parental care to all organisms which produce gametes from limited resources. In this situation, the more offspring (n) that are produced, the smaller they will be (offspring size $= z$). Moreover, offspring survival (S_j) is likely to reduce with z so that S_j and n trade-off in the required way.

Fig. 4.8 is a simple model (from Kolding & Fenchel, 1981) based on these considerations. It is assumed that poorer trophic conditions

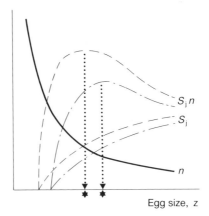

Fig. 4.8. Effect of egg size on fecundity and juvenile survival. Fitness is proportional to $S_j n$. The star shows the optimum size. If S_j is reduced (as might happen under poor trophic conditions) natural selection should favour bigger eggs. Redrawn from Kolding & Fenchel (1981).

lead to a reduction in survival for eggs of a particular size (because they have less resources to support them) as shown in Fig. 4.8. Maximum $S_j n$ then shifts to the right as shown in Fig. 4.8 and the predicted optimum egg size increases. Hence, bigger eggs should be produced in poor trophic conditions. In fact, this prediction also emerges from another trade-off (t_j vs. n) which will be considered further below (pp. 84).

A more complex, but related argument has been applied to an explanation of the reproductive strategies of marine benthic invertebrates. Here there is a discontinuity between: (a) forms which produce a large number of small eggs with **planktotrophic** (indirect) development and (b) those which produce a smaller number of large eggs with **lecithotrophic** (or direct) development. This has been considered in terms of the trade-off between n and juvenile survival. The latter reduces with time in the plankton. Models of this situation were initiated by Vance (1973a & b) and extended by Christiansen & Fenchel (1979). Both assume constant reproductive investment so that n is related negatively to z. The predictions depend on (1) risks experienced in the plankton and (2) the speed with which this stage can be completed and hence on trophic conditions. With biologically reasonable assumptions about these the models predict that poor conditions for growth in the plankton will favour direct development from larger eggs (or lecithotrophy) and this may explain why the percentage of species with lecithotrophic larvae increases with latitude and with depth. More-over, the models also predict that only two evolutionarily stable states are likely and hence explain the discontinuity in life-cycle types noted above.

However, $S_j n$ is a good measure of fitness only if t_j is unaffected by z and n, which is reasonable for seasonal breeders where t_j is fixed but not for non-seasonal breeders where t_j can change. The trade-off between t_j and n is considered below. Interestingly the predictions that emerge from this latter analysis are also compatible with the data discussed above.

4.6 Costs of parental care
$(S_j$ vs. S_a; t_j vs. S_a; t_j vs. t_a; S_j vs. $t_a)$

Certain animals make a metabolic investment in caring for eggs and/or juveniles after they are released. Four possible trade-offs are involved and these are summarized in the subheading.

Whenever a parent takes risks to defend its offspring or to provide resources for their development, S_a is being traded-off against S_j and t_j respectively. Given information on the form of the trade-offs, optimum solutions can be found (Sibly & Calow, 1983). The trade-offs will usually be 'behavioural', possibly involving active defense of the off-spring or their feeding grounds. For example, parent swans and geese actively defend a feeding space around their offspring allowing them to feed unhindered (Owen, 1980). However, physiological trade-offs can also be involved. For example, brood ventilation by aquatic animals might increase the supply of oxygen to eggs and broodlings and there-by improve their chances of successful and rapid development, particu-larly under conditions of hypoxia. This is thought to be of relevance in the ecology of some aquatic leeches (Calow & Riley, 1982), brooding octopuses (Calow, in press) and sticklebacks (Wootton, 1984).

On the relationship between S_j and S_a, it can reasonably be assumed that if the parent sacrifices nothing, few or no offspring will survive. As more sacrifices are made by the parent (S_a reduces), juvenile survival will be improved, but it is unlikely that the negative relationship between S_j and S_a will be linear. A plausible sigmoid relationship is proposed in Fig. 4.9 with S_j improving slightly with the initial reduc-tions in S_a and, no matter how much effort is invested by the parent, never reaching a juvenile survivorship of 1. Also included in Fig. 4.9 is a coordinate representing the extent to which the offspring are threat-ened by environmental conditions (e.g. low oxygen tension). Clearly as the threat reduces (e.g. oxygen tension increases) S_j improves and the effect of the parent, and hence of sacrifices in S_a, on S_j reduces to zero.

Another kind of parental care involves parents taking risks so that their offspring can develop more rapidly, so t_j is an increasing function of S_a. A simple plausible function relating these two variables is the concave one shown in Fig. 4.10a. Note that it is assumed that t_j will always be greater than zero no matter how many risks are taken. Also illustrated in this graph is the effect of the intensity of an environmental threat to the offspring. Again, as this reduces, t_j reduces and becomes less sensitive to S_a.

The optimum investment by parents in these kinds of care — i.e. the extent to which a parent should allow its metabolic or behavioural investment to depress S_a — can be found for the trade-offs that have been postulated (Sibly & Calow, 1983; Calow, 1984). For the trade-off involving S_a and S_j it occurs somewhere on the descending part of the sigmoid curve (maximum S_j obtained for minimum reduction in S_a —

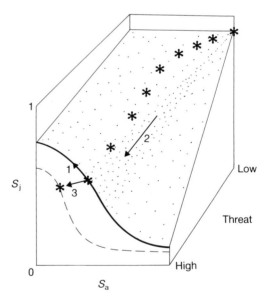

Fig. 4.9. Relation between juvenile survival (S_j) and survival rates of a caring parent (S_a) and effect of a threat on optimum parental risk. More risk should be taken with higher fecundity (1), greater threat (2) and with increasing sensitivity of juveniles to a particular threat (3). It should be borne in mind, however, that we know little about the actual shape of this surface, and that the Principle of Compensation (Chapter 1) may introduce complications.

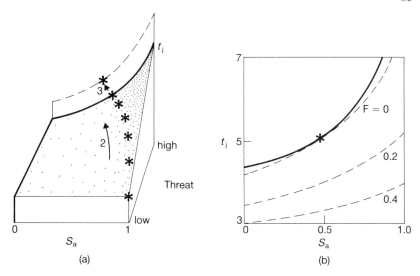

Fig. 4.10. Relationship between developmental time, t_j, and survivorship of a caring parent (S_a). F-contours are shown as broken lines in (b) and increase towards bottom right (calculated for $n = 100$, $\mu_j = 0.92$). All other conventions (and caveats) as in Fig. 4.9.

see Fig. 4.9). Moreover, it can also be shown from this analysis that provided S_j is a monotonic, decreasing function of S_a, then optimal S_a is smaller for higher fecundity (n). Thus, other things being equal, a parent should take more risks in parental care as brood size increases, and this seems to be true for male three spine sticklebacks (Pressley, 1981) and female redwing blackbirds (Robertson & Bierman, 1979). It is also clear from the graphs that investment should increase with the intensity of the threat from the environment and also with the increasing sensitivity of the juveniles to the threat. Finally, the effectiveness of the parental care (in terms of its affect on S_j) will also influence the optimal investment but this might alter the trade-off in one of several ways so that an unambiguous prediction is not possible.

Fitness contours can also be determined on the S_a–t_j landscape and are shown as broken lines in Fig. 4.10b. The optimal solutions are the points on the trade-off curves that make contact with the contour furthest to the right. Again these will depend on the level of the threat and the sensitivity of the juveniles.

Note, however, that these models make no allowance for possible genetic differences between parents and offspring that might lead to parent-offspring conflicts. These complicate the optimization arguments and have been considered by Trivers (1974). Consider the fate of a conflictor gene which enables juveniles carrying the gene to attract extra investment from their parents. When such genes are rare in the population they occur, according to Mendel's Laws, on average in half the offspring. Conflictor genes will spread in the population if the benefits they obtain outweigh the costs incurred because of the selfish behaviour of their conflictor sibs (remember that at this stage half the sibs carry the conflictor gene). However as the conflictor gene becomes common, most of the sibs will carry it, so the cost/benefit ratio worsens, and the stage is set for the evolution of parent-offspring conflict (Charnov, 1982; Parker, 1985).

These complications apart, the models in this section can make a number of at least qualitative predictions which can be tested. The arguments are, nevertheless, speculative and what is now required is (a) information on the form of the trade-offs and how they are influenced by relevant environmental variables and (b) crucial tests of the predictions that emerge from the models.

4.7 Trade-off between fecundity and time to first breeding (n vs. t_j)

If a fixed amount of resource is available for producing and provisioning gametes (say P_r), then a smaller number of gametes (n) will be produced if each is allocated more resources (z):

$$z = \frac{P_r}{n} \qquad (4.7)$$

Moreover, developmental time t_j (from gamete to adult) is likely to reduce with z and so there will be trade-offs between n and z and hence n and t_i, as in Fig. 4.11.

Returning to equation 4.2, ($1 = \frac{1}{2}e^{-(F+\mu_i)t_i} n + e^{-(F+\mu_a)t_a}$) this can be manipulated to show that on plots of t_j vs. $\log_e z$ contours representing equal fitness will be straight lines with negative slope (Box 4.3). To appreciate this, consider Fig. 4.12a. As you move to the left along a fixed value of t_j (dotted line), z decreases so n increases and this is positively correlated with F. Alternatively if you move *down* a vertical line corresponding to a fixed value of z, (broken line) t_j reduces and F increases.

We can get some idea of the trade-offs operating in these landscapes by imagining possible relationships between size and developmental time in organisms. These may be deducible from growth curves, which are often sigmoid as in Fig. 4.13a. Switching the axes and making a simplifying assumption about growth one obtains the trade-off curves as in Fig. 4.13b and c. Optima are the points on these curves which make contact with the highest contours of Fig. 4.12b.

If the influence of environmental factors on the shape and position of these curves can be specified, then we can make predictions about the evolution of the size of reproductive propagules. For example, as conditions become poorer for individual growth the trade-off curve is likely to move 'up' (since it will take longer for a propagule of a particular size and hence associated with a particular n to develop) and become steeper (since small propagules will be affected proportionately more than large — see Fig. 4.11 for proof of this). Hence, the larger propagules should be produced as conditions for growth become poorer. This prediction also emerged out of an earlier analysis involving juvenile survival (S_j) pp 79–80 and therefore, ought to be widely applicable. Observations relating to this will be considered further in Chapter 7. The Principle of Compensation is unlikely to affect these trade-offs since compensation can occur through adjustment of either t_a or μ_a without any effect on the optimal strategy.

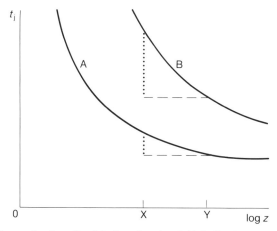

Fig. 4.11. Time to first breeding (t_j) plotted against initial offspring size ($\log_e$ scale), z. Curves A and B represent good and poor growth conditions for offspring. Dotted lines indicate times taken to grow from size X to size Y. Note that it takes longer to grow from size X to Y in environment B — this is because growth conditions are worse there. It follows that the trade-off curve is steeper where growth conditions are worse.

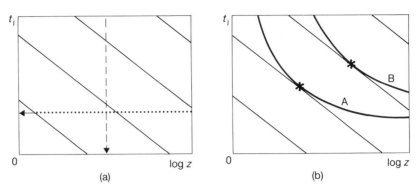

Fig. 4.12. Diagonal lines represent fitness contours. *F* increases in direction of arrows in (a) — see text. (b) shows the two trade-off curves in Fig. 4.11 with their associated optimal strategies (starred).

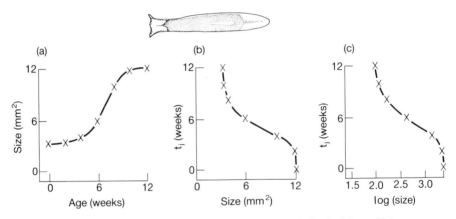

Fig. 4.13. (a) Growth curves for the freshwater flatworm *Polycelis felina*, which reproduces by binary fission. From Calow *et al.* (1979). (b) Switching axes gives the trade-off between size and t_j, provided one makes the assumption that juveniles grown to a particular size take as long to reproduce as other fission products starting at that size. (c) Log_e transform of the horizontal scale gives the trade-off in the form of Fig. 4.11.

4.8   Complications of direct costing

We now return to the complications of the direct-costing model. We have analysed this more thoroughly elsewhere (Sibly & Calow, 1984) where, on the basis of a number of simplifying assumptions about the shape of trade-off curves and the value of *F*, we show that:

1   Optima are unaffected if trade-offs are between fitness components

---

BOX 4.3. Trade-off between n and t_j.

Let $t_j(z)$ be the trade-off curve between t_j and egg size z, as illustrated in Fig. 4.11. Writing $x = \log_e z$ and substituting in equation 4.2 gives

$$1 = \tfrac{1}{2}e^{-(F+\mu_j)t_j - x}(P_r) + e^{-(F+\mu_a)t_a} \qquad (4.8)$$

which can be used to draw fitness contours when only t_j and x are allowed to vary. If everything but t_j and x is constant in equation 4.8 then $(F + \mu_j)\, t_j + x =$ constant, so that on a plot of x vs. t_j, fitness contours are straight lines with equation $t_j = -x/(F + \mu_j) +$ constant. Hence the slope of these fitness contours is $-1/(F + \mu_j)$.

---

of the same age classes; i.e. all the trade-offs in the next chapter and also the n vs. t_j and n vs S_j trade-offs of this one.

2 In other cases, where parental components are traded against offspring (the stuff of this chapter), the direct-costing optimum in general involves a smaller investment by the parent in producing or caring for offspring. The reasons for this are intuitively clear: direct-accounting costs, reducing S_a or increasing t_a (Fig. 4.1b), are incurred before the release of the first propagules. Hence, high costs here might mean that no viable propagules (gametes or broodlings) are released at all. However, this conclusion can be complicated by the Principle of Compensation, as discussion in Chapter 1 (p. 21), so that predictions from the direct-costing model are not straightforward.

4.9 Conclusions and summary

The above analyses have again emphasized that physiological trade-offs can play a crucial part in the process of adaptation. At least 10 trade-offs are possible between the components of the Euler/Lotka equation (Table 4.1) and which ones are important in particular instances will depend upon both the organism involved and its environment.

The simplest hypothesis is that there are no trade-offs and in this case it is expected that life cycles would maximize S and n and minimize t irrespective of accounting methods. But this is so unlikely that it is probably not worth contemplating. On the other hand, the n vs. S_a trade-off is likely to be very general, because the number of offspring produced per breeding season is often likely to have some impact on the survivorship of the adult between breeding seasons (but cf. Bell, 1984a & b). If parental care occurs, a number of other trade-offs become important and these are discussed on p. 81. If $t_a = t_j =$ one year, as is true of many temperate species, then trade-offs involving developmental time and t_a will be unimportant since the need to conform to the annual environmental cycle will be paramount. If, however, t_a or $t_j \neq$ one year, trade-offs involving developmental rates become very important. The trade-off n vs. t_j is likely to apply very widely since if more offspring are produced it will generally take longer for each to develop to a size and state at which reproduction is possible.

How sensitive are the conclusions to the simplifying assumptions made in the derivation? One assumption was that age differences in breeding adults (i.e. in S_a, t_a and n) can be ignored. A number of authors have considered the implications of specified age- or size-dependent changes in breeding performance (reviewed in Charlesworth, 1980). However, a major difficulty is that these predictions depend crucially upon the assumed age- or size-dependent changes and these are difficult to define *a priori* (see Chapter 6).

Another important assumption was that life-cycle variables and trade-offs are of a deterministic kind. In the real world, however, they are likely to change in space and time. Schaffer (1974) considered a situation where the environment alternated between good and bad conditions and where survivorship and fecundity altered as a result of this variation; increasing by a constant factor in good years and reducing by a constant amount in poor years. A major outcome from this analysis was the so-called **'bet-hedging' hypothesis** — which predicts that if fecundity or juvenile survivorship varies, adults should invest less in reproduction than expected from the deterministic models (see also Goodman, 1984). However if adult survivorship varies adults should invest more. Schaffer (1974) considered the case where $t_j = t_a = 1$, so $e^F = n\,S_j + S_a$. He supposed that environmental variation caused $n\,S_j$ and S_a to vary by an *amount proportional to their mean values*; i.e. $(nS_j) \pm (nS_j)x$ or $S_a \pm S_a y$.

However, the environmental variation need not operate in this way; it could cause (nS_j) to change by an amount which did not depend upon (nS_j) e.g. from (nS_j) to $(nS_j) \pm x$. Now the optimal strategy is the same as in the deterministic case.

5
Growth

5.1 Introduction

In the course of development, resources obtained from food and not used to pay costs of living are invested first in somatic and then reproductive biomass (Chapter 3). It was convenient to treat reproduction separately in the last chapter, but this avoided the issues of when the switch should occur (i.e. how big should animals become before they start reproducing?) and how the switch should be effected (sharply or progressively). This chapter addresses these questions and considers patterns of somatic growth in general. Questions of size are addressed further in the next chapter.

5.2 Metabolic framework

Metabolic conditions allow growth to take place when the input of nutrients exceeds the catabolic output (respiratory heat losses + organic waste; Chapter 3). If the functional relationship between body size and rates of both feeding and catabolism are known, then growth patterns and rates can be predicted. This is the basis of the von Bertalanffy (1960) growth equation. Suppose that rates of feeding and catabolism both scale allometrically with body size (cf. Chapter 6) such that

$$\text{Food input} = I = am^b \qquad (5.1)$$
$$\text{Catabolism} = C = cm^d \qquad (5.2)$$

for appropriate constants a to d. Then growth rate is given by

$$dm/dt = am^b - cm^d \qquad (5.3)$$

Taking logarithms of equations 5.1 and 5.2 gives

$$\text{Log } I = \log a + b \log m \qquad (5.4)$$
$$\text{Log } C = \log c + d \log m \qquad (5.5)$$

There are 3 possible relationships between b and d:

1 b < d
2 b = d
3 b > d

and these are illustrated in Fig 5.1a. Their implications for the 'growth curve' (how size changes with age) are best understood by considering how the surplus $I - C$ ($=P$) changes with m (Fig. 5.1b). Since I and C are rates, so is P and is equivalent to growth rate (dm/dt). Hence, from the relationships in Fig. 5.1b we can arrive at the appropriate growth curve (Fig. 5.1c). With case 1 the input and cost curves converge in Fig. 5.1a. Because of the logarithmic relationship, P increases to a maximum with m before reducing to zero (point of intersection of 2 curves). Hence size increases at an increasing rate with age to the point where P is maximum and then there is an inflexion and size increases at a reducing rate on to a maximum size when the curves for I and C intersect and $P = 0$. This results in a sigmoid growth curve. In the other 2 cases, P increases increasingly with size and so m increases increasingly with age.

Sigmoid curves are the most common types of growth curve in metazoan animals (e.g. Fig. 5.2, Fig. 4.13a). However, continuously rising growth curves, with a sharp cut-off (J-curves) have been recorded for some organisms which use body surfaces for both the absorption of food and the uptake of oxygen (e.g. endoparasites; Calow, 1981). Even here, though, mechanical constraints can lead to a sigmoid growth curve.

5.3   Cellular framework

Except for a few taxa (e.g. nematodes and rotifers), growth in animals occurs mainly by the proliferation of cells rather than by the hypertrophy of individual cells. That is to say, P is used to form more cells rather than to expand existing cells. Cell division is, therefore, at the heart of the growth process. Cells formed in the process become specialized in form and function in the production of different tissues, a process known as **differentiation**. For reasons that are not fully understood, cell division and differentiation are often mutually exclusive so that the accumulation of differentiated tissue within organisms probably also sets limits to rates and patterns of growth.

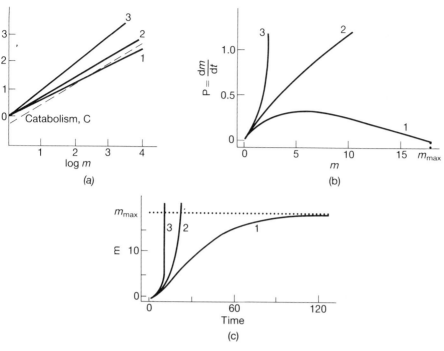

Fig. 5.1. Three different growth curves, labelled 1, 2, and 3, showing their dependence on rates of feeding (I) and catabolism (C). (a) shows a log-log plot of I against bodyweight (m) for three different allometric relationships of food input to weight: (1) $I = m^{0.67}$ (2) $I = m^{0.75}$ (3) $I = m^{1}$. Also shown is the rate of catabolism C (broken line), assumed to follow $C = 0.78\,m^{0.75}$. (b) shows how $P\ (= I - C = \dfrac{dm}{dt}$ = growth rate) varies with weight. Note that for curve 1 a maximum size, m_{max}, will be reached when $\dfrac{dm}{dt} = 0$. (c) shows the resulting growth curves, obtained by numerical integration. 1 is a sigmoid growth curve.

Weiss & Kavanau (1957) proposed that these limitations were based upon a balance between growth promoting and inhibiting factors and incorporated their ideas into a model that generated realistic growth curves. This is summarized in Box 5.1. Another possibility is that the formation of complex tissues and organs might take more time than simple mitotic division and impose a bottleneck on development. For example, Sacher and Staffeldt (1974) have discovered a good positive correlation between the period of gestation and neonatal brain size in mammals and have proposed that brain development is a bottleneck in foetal ontogeny (but cf. Case 1978, and Eisenberg, 1981).

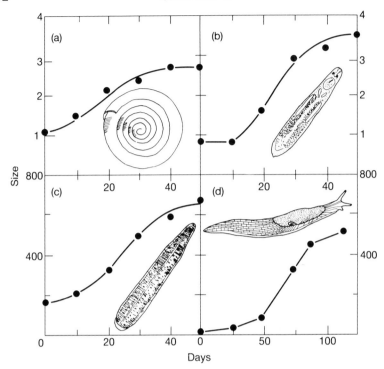

Fig. 5.2. Sigmoid growth curves of several invertebrates: (a) shell diameter (mm) of a
planorbid snail; (b) body plan area (mm²) of a rhabdocoel flatworm; (c) joules
potential energy of an erpobdellid leech; (d) mg fresh weight of a slug. After Calow
(1981).

5.4   Involvement of selection

The metabolic and cellular models are complementary and provide a
framework of constraints within which natural selection can operate. If
there were no fitness costs involved, selection would maximize growth
rates within these constraints because this would minimize development
time and possibly maximize survival. This is an explanation favoured
by Ricklefs (1984). Yet we have already noted several lines of evidence
which suggest that growth rates are not always at maximum levels
(Chapter 3). This suggests that there are costs of growing fast in the
juvenile phase and these can only be mortality costs — a suggestion
implicit in Lack (1968) and Williams (1966) but first explored rigorously
by Case (1978). Possibilities, though admittedly without much experimental foundation (Ricklefs, 1984), are reviewed in Chapter 3.

  Before the first breeding season, it is appropriate to express the

BOX 5.1. Weiss and Kavanau model

1 Tissue is thought to consist of generative mass
(GM) which divides to produce more GM and non-
dividing, differentiated mass (DM) (Fig. 5.3a).

2 GM produces a template molecule (●) which sti-
mulates growth but which cannot escape from cells,
and an antitemplate (△) which inhibits ● and which
can escape into the intercellular space (ICP). When
● becomes inhibited by △, GM transforms to DM.
(Note both ● and △ are hypothetical).

3 Rate of production of GM depends on amount of
GM so

$$dGM/dt = a \ GM$$

where a is a constant. This predicts exponential
growth (Fig. 5.3b curve 1).

4 Rate of conversion of GM to DM is also likely to
depend on GM, i.e. $dGM/dt = a \ GM - b \ GM$
which yields curve 2.

5 The negative feedback effect of △ is modelled by
assuming that it is proportional to x^y where x =
concentration of △, and y > 1, so

$$dGM/dt = a \ GM - b \ GM - c \ x^y$$

which yields curve 3.

6 Finally catabolism and excretion of the breakdown
products of GM, DM and △ are taken into account
as follows

$$dGM/dt = a \ GM - b \ GM - c \ x^y - \text{Excretion}$$

which yields curve 4.

Note: *Curves 1 and 2 increase continuously with time;
3 and 4 are sigmoid.*

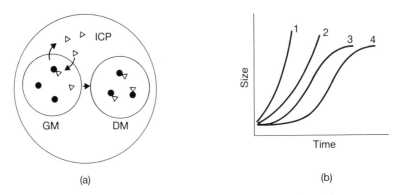

Fig. 5.3. A pictorial representation of the Weiss-Kavanau growth model

fitness of a trait in terms of its effects on the risks of mortality (μ_j) and age at first breeding (t_j) and arguments in this section are framed accordingly. They are mainly from Sibly *et al.* (1985).

There are two components of the growth curve that can influence μ_j: (1) size (m) at a given time/age and (2) the rate of growth $dm/dt = \dot{m}$ associated with organisms of that size, so formally:

$$\mu_j = \mu_j \ (m, \ \dot{m}) \tag{5.6}$$

$$\dot{m} \text{ is related to } m \text{ by } \dot{m} = mu \tag{5.7}$$

where u is the rate of growth per unit weight, with a maximum value K. u will be considered to be the control variable (p. 20); i.e. the quantity moulded by natural selection.

The simplest possible relationships between μ_j, m and u are:

Assumption   5.1a μ_j is a monotonic decreasing function of m (Fig. 5.4a)
5.1b μ_j is independent of m (Fig. 5.4b)
5.1c μ_j is a monotonic increasing function of m (Fig. 5.4c)

Assumption   5.2a μ_j is a convex function of u viewed from below (Fig. 5.5a)
5.2b μ_j is a straight line function of u (Fig. 5.5b)
5.2c μ_j is a concave function of u viewed from below (Fig. 5.5c)

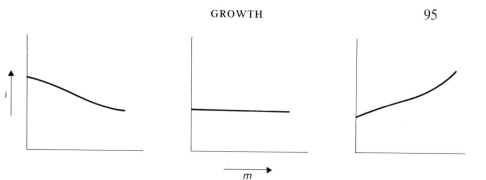

Fig. 5.4. Possible forms of $\mu_i(m)$.

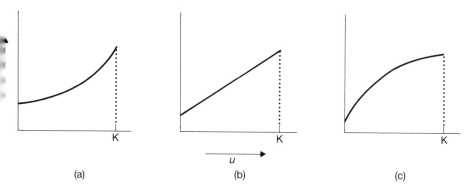

Fig. 5.5. Possible forms of $\mu_j(u)$. u has a minimum value of 0 and a maximum of K.

Note: μ_j is almost certainly an increasing function of u (Chapter 4), though the mathematics below does not require this assumption.

The evolutionary problem is to choose values of growth rate $u(t)$ for each moment of time so that fitness will be maximized. Because growing as fast as possible ($u=K$) subjects the animal to maximum mortality (Chapter 4) it is not necessarily the best thing to do. The benefits of growing faster (maturing earlier) have to be weighed against the mortality costs. Because $u(t)$ can vary with time, we are dealing with a dynamic model, and the optimal strategy can only be found by Pontryagin's method (Box 1.4). A suitable measure of fitness for this

situation is $\Phi = -\int_0^{t_j} (F + \mu_j(m,\dot{m}))\, dt$ (Sibly *et al.* 1985). Formally

the problem is to choose $u(t)$ and t_j so as to maximize Φ. In this problem we suppose that initial size (at birth) and final size (at maturity) are given and that there can be no reverse growth. In the language of Box 1.4 the initial and final values of the state variable (m) are fixed but terminal time (t_j) is free. Plugging in the mathematics of Box 1.4 reveals that the optimal strategy chooses values of growth rate $u(t)$ to maximize the Hamiltonian, H, defined by

$$H = -F - \mu_j(m, u) + \lambda m u \qquad (5.8)$$

This can be rewritten as

$$H = -\mu_j + \lambda m u - F \qquad (5.9)$$

Thus contours of equal H in the $\mu_j - u$ plane are straight lines with slope λm, (i.e. to keep H constant, u must increase in proportion to μ, with λm representing the constant of proportionality). Contours with bigger H values occur lower down the μ_j axis (example in Fig. 5.6).

It follows that if $\mu_j(u)$ is concave viewed from below (Assumption 5.2c) then the optimal strategy (i.e. that which maximizes H, which is equivalent to maximizing fitness — see Box 1.4) is either $u = 0$ or $u = K$ and since $u = 0$ is a biological nonsense it follows that $u = K$ throughout growth. To find how $\log_e$ (body mass) changes with time note that

$$\frac{d\,(\log_e \text{ body mass})}{dt} = \frac{1}{m}\frac{dm}{dt} = u, \text{ by equation 5.7} \qquad (5.10)$$

In the present case $u = K$ and therefore $\log_e$ (mass) increases linearly, with slope K (Fig. 5.7 curve i). The same applies if μ_j is a straight line function of u (Assumption 5.2b) (except in the unlikely event that the slope of the H-line equals that of $\mu_j(u)$ in which case all values of u are equally good).

When $\mu_j(u)$ is convex viewed from below (example in Fig. 5.6a) intermediate strategies ($u < K$) may be optimal. Two possibilities, with λm positive, are illustrated in Figures 5.6b and 5.6c, from which it can be seen that which strategy is optimal depends on the slope of the H-lines and hence on λm. The way λm is influenced by assumptions la to lc can be deduced algebraically (Sibly et al., 1985).

The optimal strategies in the different cases covered by Assumptions 5.1a–c and 5.2a–c are summarized in Table 5.1 and a worked example is given in Box 5.2.

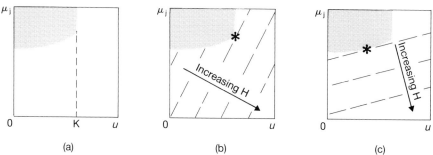

(a)   (b)   (c)

Fig. 5.6. (a) An animal's growth options (shaded), bounded below by the trade-off curve in Fig. 5.5a, and to the right by the constraint $u = K$. (b) and (c) The broken lines are lines of equal H, and H increases as one descends the μ_j axis, as indicated by the arrows. Asterisks denote optimal strategies, i.e. those points in the options sets which lie on the H-line corresponding to the highest possible value of H.

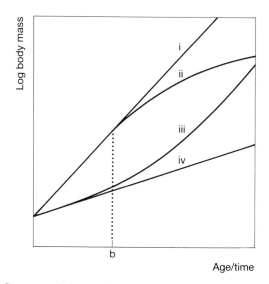

Fig. 5.7. Curve types i—iv as referred to in the text and in Table 5.1. b marks the end of the first phase (of linear growth) for curve ii.

Animal growth curves are generally found to be either sigmoid or positive exponential (p. 90). Translated onto a log-linear plot, as in Fig. 5.7, these are type ii or linear respectively. No examples of type iii curves are known. Referring to Table 5.1 we see that type ii curves are optimal if Assumptions 5.1a and 5.2a are true, and type iii curves are optimal if 5.1c and 5.2a are true. Since type ii curves exist in nature

Table 5.1. Optimal values of u during growth with curve types as in Fig. 5.7 in brackets

		Assumption 5.2a $\mu(u)$ convex viewed from below	5.2b straight	5.2c concave viewed from below
Assumption	$\mu(m)$			
5.1a	Decreasing	Decreasing (ii)	Maximum (i)	Maximum (i)
5.1b	Horizontal	Constant (iv)	Maximum (i)	Maximum (i)
5.1c	Increasing	Increasing (iii)	Maximum (i)	Maximum (i)

but type iii curves do not it is interesting to observe that where natural demographies are known they generally follow Assumption 5.1a; i.e. survivorship curves suggest that mortality risks reduce with body size in natural populations, most frequently at a reducing rate, reaching a minimum at the age of first breeding (Fig. 5.8, Itô, 1980). The reverse, increasing mortality risks with size, is usually a result of physiological malfunctions associated with ageing processes which will be ignored here and anyway are rare for all but domesticated and laboratory animals and humans.

There is little information on the relationship between mortality rate (μ_j) and growth rate (u) but there are grounds for believing that μ_j ought to be an increasing function of u, convex when viewed from below (Assumption 5.2a). It would seem likely, for example, that the chance of developmental mistakes and the risks of stress increase increasingly as the rate of construction of the organism increases, i.e. it is possible that a system of physiological priorities operates so that resources are first diverted to growth from those maintenance processes that have little effect on μ, and are only removed from very important maintenance processes when all else has been used (see also p. 59). Even in species in which offspring enjoy parental care, high growth rates will require high brooding investments by parents and these will entail increased risks thus putting the whole family group in jeopardy (Case, 1978). Though it should not be taken too seriously, the inter-specific comparison illustrated in Fig. 3.9 is compatible with this relationship.

If Assumption 5.2a is true (though not otherwise) animals should 'hurry through' the vulnerable stages of their life cycles. This conclusion was arrived at intuitively by Lack (1968) and Williams (1966). More specifically, growth rates should increase with vulnerability, so that

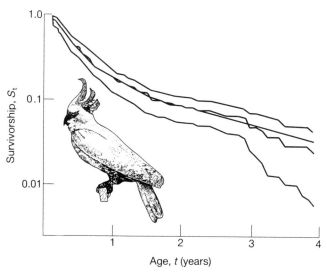

Fig. 5.8. Survivorship curve, with 95% confidence interval, for *Cacatua roseicapilla*, a species of cockatoo. Note that the curve is steep initially but gets gradually shallower. The slope of the curve is equal to mortality rate. This suggests that mortality rate decreases with age, reaching a minimum (constant slope) at age 2, which is the age of first breeding. From Buckland (1982).

species suffering high offspring mortality should have higher growth rates than those with lower mortality levels. Ricklefs (1969) tested this prediction for birds, but could find no significant correlation between mortality and growth rate. However, his analysis is open to the criticism that mortality rates were not defined very precisely and since mortality rates vary considerably between years, reported values cannot be taken too seriously (Case, 1978; Ricklefs, 1984). Comparisons between birds whose habitat and nesting habits are so different that large differences exist between species in infant mortality and growth rates, leave little doubt that species habitually exposed to nest predators fledge their young more quickly than less exposed species (Lack, 1968; Cody, 1973). Similarly, in mammals, high growth rates are often associated with high infant mortality. For example, the infant mortality of the generally fast growing artiodactyls is extremely high whereas elephants, with slow growth, have low infant mortality (see Case, 1978 who cites more examples).

Note that we assumed that K is the maximum growth rate feasible for the organism, and this is expected to be attained in the initial phase of growth. Hence a specific prediction is that it should not be possible to enhance growth rates over this this period e.g. by hormone treatment. Failure of this assumption would have little effect on predictions if Assumption 5.2a were true, except that this 'first phase' of growth in Type ii curves would disappear. However, failure of this assumption if Assumpions 5.2b or 5.c were true, would lead to the prediction of infinite growth rates, a biological nonsense.

BOX 5.2.

As an example suppose that

$$\mu_j = 0.1 + \frac{u^2}{100} + \frac{1}{m^4}, \tag{5.11}$$

where $m(0) = 1$, $m(t_j) = 7.07$, constraint $K = 0.427$, and $F = -0.1$. From equation (5.8),

$$H = 0.1 - 0.1 - \frac{u^2}{100} - \frac{1}{m^4} + yu \tag{5.12}$$

where $y = \lambda m$. $\tag{5.13}$

$\mu_j(u)$ is convex viewed from below, so Assumption 5.2a holds, and μ_j is a decreasing function of m so Assumption 5.1a holds.

$$\text{Since } y = \lambda m, \quad \frac{dy}{dt} = \frac{d\lambda}{dt} m + \lambda \frac{dm}{dt}, \tag{5.14}$$

and substituting for $\frac{d\lambda}{dt}$ and $\frac{dm}{dt}$ from equations (1.15)

and (5.7) respectively we obtain

$$\frac{dy}{dt} = -\frac{4}{m^5} m = -\frac{4}{m^4}. \tag{5.15}$$

Initially the constraint $u = K$ may be binding (see above), in which case

$$m(t) = e^{0.427t}, \text{ using equation (5.10)}. \tag{5.16}$$

Thereafter an intermediate strategy is optimal so that

$\frac{\partial H}{\partial u} = 0$, i.e. $-\frac{u}{50} + y = 0$ by differentiation of equation (5.12).

$$\therefore \frac{dy}{dt} = \frac{1}{50}\frac{du}{dt}. \qquad (5.17)$$

$$\text{Let } \log_e m = x. \qquad (5.18)$$

Then $u = \dfrac{dx}{dt}$ by equation (5.10), $\qquad$ (5.19)

and $\dfrac{du}{dt} = \dfrac{d^2 x}{dt^2}.$ $\qquad$ (5.20)

But $\dfrac{du}{dt} = 50\dfrac{dy}{dt}$, by equation (5.17),

$$= -\frac{50 \times 4}{m^4}\text{ by equation (5.15)},$$

$$= -\frac{200}{e^{4x}}\text{ by equation (5.18)},$$

$$\therefore \frac{d^2 x}{dt^2} = -200\, e^{-4x}. \qquad (5.21)$$

It can be verified that this is satisfied by

$$x = \tfrac{1}{2}\log_e (20t - c), \qquad (5.22)$$

for some constant c to be determined. Although this is not the general solution to equation (5.21) it allows us to obtain the optimal strategy (see below). By equation (5.19)

$$u = \frac{10}{20t - c}. \qquad (5.23)$$

The second phase of growth starts when the constraint $u = K$ is just binding. Let this moment be $t = t'$. $\qquad$ (5.24)

$$\text{Then } u = 0.427 = \frac{10}{20t' - c}, \qquad (5.25)$$

and size is then given both by equations (5.22) and (5.16), which can only hold simultaneously if c = 50 and $t' = 3.67$. Hence size increases exponentially according to equation (5.16) from $t = 0$ to $t = 3.67$ and thereafter according to:

$$m = (20t - 50)^{1/2} \text{ (from equations (5.18) and}$$
$$(5.22)). \qquad (5.26)$$

u is bound by the constraint $u = 0.427$ until $t = 3.67$, and thereafter

$$u = \frac{10}{20t - 50}, \text{ from equation (5.23).} \quad (5.27)$$

Terminal size $m = 7.07$ is attained at $t = 5$ which is, therefore, terminal time. It can be shown that $H = 0$ at terminal time so that $H(t_j) = 0$ as required by equation (1.17). The growth pattern is therefore sigmoid as illustrated in Figure 5.9.

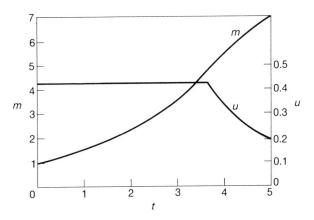

Fig. 5.9. A sigmoid growth curve plotting bodyweight m vs. age t derived from the risk-taking model of Box 5.2. Exponential growth occurs during the first phase of growth, until $t = 3.67$, and during this period the constraint $K = 0.427$ is binding on u. Thereafter, in the second phase of growth, u decreases according to equation (5.27).

5.5   Catch-up growth

Above-normal growth rates (for size or age) sometimes occur after periods of growth retardation or inhibition. For example, Fig. 5.10 shows data collected by Stebbing (1981) on the growth of the colonial hydroid *Campanularia flexuosa* exposed to a range of copper concentrations. Copper inhibits growth, particularly at levels greater than 15 g l^{-1}. After a return to normal sea water there is a spurt of growth in these animals. Such catch-up phenomena occur in human children following illness and starvation (Prader *et al.*, 1963) and have also been described in rats and chickens (Jackson, 1937; von Bertalanffy, 1960;

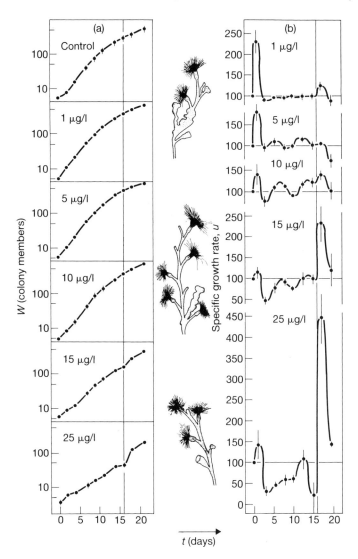

Fig. 5.10. Growth of a colonial hydroid exposed to a range of copper concentrations for 16 days, after which the colony was returned to sea water without added copper. (a) Cumulative curves of colony members. (b) Specific growth rates. Data points are the mean of five colonies and bars indicate standard errors of the means. After Stebbing (1981).

McCance & Widowson, 1962) and some invertebrates (Calow, 1981). They are often referred to as being adaptive — i.e. catching up with the normal age-specific size may improve fitness. Here we consider if this interpretation is consistent with the optimization analysis.

In considering this question, it is important to note that optimization analyses are prospective (i.e. related to the *present* state of organisms and the *future* environment they face). Natural selection is also prospective in the sense that it will act on the state of organisms in relation to the environments they later encounter. Past performance of an organism is only relevant to the future optimal strategy in so far as it affects the present state of the organism. For example, a decision on whether to continue offering parental care to a brood depends on expected future fitness returns from so doing and these depend on the present state of the parents and brood, and expected future environmental conditions. If, given these, the best option is to cease brooding, that option should be taken, no matter how much has previously been invested in brooding (Dawkins & Carlisle, 1976).

Thus, in a prospective analysis of growth, the optimum strategy depends upon: (1) the initial state of the animal and (2) the future environment. If, after a period of growth inhibition, (2) remains the same, the optimal strategy depends only on (1). By definition (1) is unchanged and so the optimum strategy, after inhibition, is the same as that before; i.e. *growth rate should proceed as normal and no catch-up is predicted.* However, the risks associated with *m* might alter through the period of inhibition; for example, predators more experienced with a particular size class might develop a 'search image' for it (Dawkins, 1971) so the cessation of growth might make that size class more vulnerable (and, in principle, the same could apply with respect to agents of disease). Moreover, the argument as stated refers only to development rate and mortality risks and yet fecundity might be a relevant parameter. Thus, if the environment changes with time, and certain environmental conditions are better for reproduction than others, catch-up might bring gains in terms of fecundity.

In conclusion, then, catch-up growth is not an unequivocally optimal strategy, but theory can make precise predictions about when it might be advantageous. Seasonality, a changing environment with certain conditions being better for reproduction than others, is likely to be the best explanation for the evolution of catch-up growth. However, this cannot be the case for non-seasonal humans and laboratory rats. Here, other factors, such as increasing risks associated with individuals retarded in size, may be important.

5.6   Growth with reproduction

For an organism of a given size (m), some of the total energy available for production ($P(m)$) must be used for reproduction at some stage. There is, in a sense, always competition between somatic growth and reproduction for a share of $P(m)$ and this section considers how that should be resolved. We will not consider here the problems of optimizing $P(m)$, instead we will assume it is given, as a function of m. The analysis is from Sibly *et al.* (1985). The notation is:

t = age

$S(t)$ = probability of an animal surviving from birth to t. $S(0) = 1$

$m(t)$ = body mass

$P(m)$ = production = total power devoted to growth and reproduction, a 'given' in this analysis. $P \geqslant 0$

$u(t)$ = fraction of P devoted to growth ($0 \leqslant u \leqslant 1$) = control variable

$n(u,m)$ = rate of giving birth by adults of age t, assumed to be uniquely determined by u and m

$\mu(u,m)$ = mortality rate assumed to be uniquely determined by u and m.

Survivorship S and mortality rate μ are necessarily linked by the equation:

$$\frac{dS}{dt} = -S\mu(u,m) \tag{5.28}$$

and growth rate $\dfrac{dm}{dt}$ depends on the fraction u of production P that is devoted to growth, thus:

$$\frac{dm}{dt} = uP(m) \tag{5.29}$$

It is assumed that the values of n and μ are uniquely determined by u and m, and the simplest possible forms of these relationships will be codified into assumptions to facilitate later analysis and discussion:

Assumption   5.3a n is a decreasing function of u, convex viewed from above (as in Fig. 5.11a)

5.3b n is a decreasing straight-line function of u (as in Fig. 5.11b)

5.3c n is a decreasing function of u, concave viewed from above (Fig. 5.11c)

Assumption 5.4a μ is convex function of u viewed from above
5.4b μ is a straight-line function of u
5.4c μ is a concave function of u viewed from above.

The evolutionary problem is to make the right allocations of resources to growth and reproduction so that fitness is maximized. What is allocated to growth (uP) cannot be devoted to reproduction ($(1-u)P$). It is not necessarily right to put everything into reproduction because fecundity n and mortality μ depend on body size m. Thus the advantages of immediate reproduction have to be weighed against enhanced future fecundity. $u(t)$ can vary in time, so we are again dealing with a dynamic model requiring Pontryagin's method of analysis (Box 1.4). A suitable measure of fitness for this situation is $\Phi = \int_0^\infty e^{-Ft}Sn\,dt$ (Box 1.4). Formally the problem is to choose $u(t)$ to maximize Φ. We suppose that initial size (at birth) is fixed, but there is no constraint on final size. We want to know the optimal strategy for as long as the animal lives. Translating into the language of Box 1.4, there are two state variables, m and S, and one control variable, u. The initial values of the state variables are fixed ($S(0) = 1$ because offspring are necessarily alive at age 0). Terminal time is infinity and the state variables are then free. Plugging in the mathematics of Box 1.4 reveals that the optimal strategy is to allocate resources to growth to maximize the Hamiltonian, H, defined by

$$\max H = e^{-Ft}Sn(u,m) + \lambda_1(t)uP(m) - \lambda_2(t)S\mu(u,m) \quad (5.30)$$

where $\lambda_1(t)$ and $\lambda_2(t)$ are functions whose biological meaning will be considered further in Chapter 8.

Rearranging equation (5.30) we can obtain

$$n = \frac{e^{Ft}}{S}(H - \lambda_1 uP(m) + \lambda_2 S\mu(u,m)) \quad (5.31)$$

This gives contours of equal H to plot on a graph of n against u. (Fig. 5.11d–f).

If Assumption 5.4b holds, so that μ is a straight-line function of u, then contours of equal H given by equation (5.31) are straight-line functions of u. Hence if the options set is concave viewed from above (Assumption 5.3c), then the optimal strategy is either $u = 0$ or $u =$ maximum but no intermediate value (i.e. a 'bang-bang' strategy in

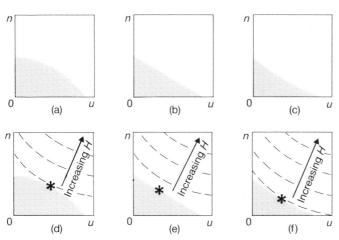

Fig. 5.11. Three possible shapes of the options set in the $n-u$ plane. (a) − (c) correspond to Assumptions 5.3a−c respectively. In (d) − (f) the broken lines are lines of equal H, as explained in the text, for a case in which μ is a concave function of u viewed from above (Assumption 5.4c), and the optimal strategies are denoted by asterisks.

engineering terminology). The same optimal strategy occurs if the options set is bounded by a straight-line function of u (Assumption 5.3b) except in the unlikely event that the slope of $n(u)$ equals the slope of the H-contours. However if the options set is convex (Assumption 5.3a) then the optimal strategy may involve intermediate values of u (i.e. $0 < u <$ maximum).

Table 5.2 summarizes the results and shows that whether the optimal strategy is bang-bang ($u =$ maximum followed by $u = 0$) or intermediate depends on the precise form of the trade-off curve $n(u)$ and the mortality curve $\mu(u)$. The bang-bang strategy is optimal in a surprising number of cases and this is at first sight puzzling since many plants and animals use an intermediate strategy; that is, they continue to grow after breeding is initiated. For example, in the Mollusca, though muricid gastropods breed only after they have ceased growing, many pulmonates and the bivalve *Mytilus edulis* grow during and between breeding (Calow, 1983b).

Our intuition is that there is a simple straight-line trade-off between reproduction and growth (i.e. Assumption 5.3b) in which case the

Table 5.2. Optimal strategies in the model of growth with reproduction. See text for details. Here convex/concave refer to curves viewed from above

	Assumption 5.3a n(u) convex	5.3b straight	5.3c concave
Assump μ (u) -tion			
5.4a Convex	Possibly intermediate	Bang-bang	Bang-bang
5.4b Straight	Probably intermediate	Bang-bang	Bang-bang
5.4c Concave	Probably intermediate	Probably intermediate	Possibly intermediate

intermediate strategy is optimal only if $\mu(u)$ is concave viewed from above (Assumption 5.4c). It would be interesting to measure $\mu(u)$ curves to see if they are concave in plants and animals which continue to grow after breeding is initiated, but this has not yet been done.

It is reasonably well-established that producing and carrying reproductive biomass is more risky than producing and carrying somatic biomass (Chapters 3 & 4). The extra risks from reproduction come from ethological/ecological causes (e.g. by making them more conspicuous and cumbersome, reproductive activities render parents more liable to accident, disease and predators) and physiological causes (e.g. by putting a burden on maintenance metabolism). It seems likely, therefore, that at least in some instances the risks arising from these increase disproportionately with investment in reproduction, (i.e. Assumption 5.4c rather than 5.4a or 5.4b) but again this has yet to be investigated precisely by experimental manipulation (Chapters 3, 4).

5.7 Conclusions and summary

Growth processes may either be maximized to physiological/developmental limits or, on the assumption that high growth rates carry survival costs, be an optimum compromise between the fitness costs and benefits of different growth rates. Unfortunately, both hypotheses can generate plausibly-shaped — in particular sigmoid — growth curves. However, according to the optimality models, sigmoid curves are not optimal for all demographic circumstances and growth rates should not be maximum at all times through the life cycle. The former is testable, but has not been tested to date and the latter is supported by some circumstantial evidence. On both counts, there is considerable scope for an experimental/observational initiative.

The optimality models can also make predictions about when animals should stop growing and start breeding. The circumstances favouring a sharp switch are broader than those favouring a progressive one. Again there is an acute need for data on the distribution of these life-cycle patterns and on the demographies and trade-offs with which they are associated.

Finally, in this and the preceeding chapter, it will have become obvious that an experimental analysis of the assumptions written into the models — e.g. on the relationship between growth rate and mortality, size and mortality, size and fecundity, etc. — is as important, if not more so, than testing the predictions themselves. This involves a research programme which balances laboratory and field, physiological and ecological approaches — something which is a central feature of the physiological ecology being advocated throughout this book.

6
Constraints of Size

6.1 Introduction

It is self evident that larger animals will eat, respire and produce more than small animals do. Hence size imposes constraints on metabolism and these are likely to be best exposed by studies on taxa with widely different sizes — i.e. interspecific comparisons. So far we have said nothing about these size constraints, though they are not only of profound physiological importance, they are also important in evaluating the adaptationist programme. For example, selection might operate simply to push physiologies to the maximum limits imposed by size. This would be compatible with the Maximization Principle (Chapter 1) but on this basis there would be a sense in which constraints (based on size limitation) could be said to be more important than the optimization of trade-offs in the evolution of physiologies.

The constraints are imposed by geometrical and physical factors, so that the form of the size-dependency can be deduced, at least in principle, *a priori* from a dimensions analysis. If the models fit the data, then the null hypothesis cannot be rejected (for a discussion of statistical methods in fitting alternative models see Smith, 1980). However, since selection might cause a downward deviation that either (i) retains the original relationship or (ii) causes a shift in the size dependence (Fig. 6.1), then allometric size dependence in itself does not argue for the primacy of constraint (cf. Stearns, 1983 1984; Reiss, 1985). In particular:

Type i means that even if the null hypothesis cannot be rejected, optimization might still be more important than constraint. It is therefore necessary to test the null hypothesis further, for example by trying to exceed the hypothesized limit by physiological (e.g. hormonal) manipulation.

Type ii means that the null hypothesis *can* be rejected and the form of this deviation might give some clues on the way natural selection influences different size-classes of organisms.

110

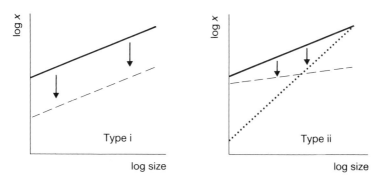

Fig. 6.1. Two types of deviation of size-dependent relationships from constraints imposed by geometrical or physical factors. The constraints are indicated as solid lines and the actual relationships as broken or dotted lines. See text for details.

In the first part of this chapter we aim to illustrate what ought to be possible in this type of programme. Our insights are still somewhat embryonic and this makes it necessary to distinguish the '*ought*' from the '*is*'. Nevertheless the field is a rapidly developing one, as witnessed by the proliferation of literature over the past few years (McMahon & Bonner, 1983; Peters, 1983; Calder, 1984; Schmidt-Nielsen, 1984; Jungers, 1985) and so we can hope for significant developments in the near future. Our aim is to use interspecific allometries to give insight into the occurrence and frequency of type ii deviations.

Given the importance of size, the second part of this chapter goes on to consider the optimal size for animals living in particular ecological circumstances. This question, about being the right size, was asked many years ago (Haldane, 1927; see also Maynard Smith, 1985) but has proved surprisingly intractable. It depends upon a variety of factors, not least being the physiological size dependencies noted above. However, of particular import are the size dependencies of fecundity and developmental rates and these are considered particularly carefully. Again for lack of good data all we can do is to outline, in general terms, the conditions that *ought* to favour an increase in size.

6.2 Size-dependency of key metabolic processes

6.2.1 CONSTRAINTS ON THE COSTS OF LIVING

The size-dependency of metabolic rate, and the hypotheses that might explain it, have been particularly well reviewed by Schmidt-Nielsen (1984) and Peters (1983), who should be consulted for further discussion of the argument presented below. Schmidt-Nielsen notes that the empirical relationship between metabolic rate and body size is known with much greater certainty for warm-blooded vertebrates than for cold-blooded vertebrates and invertebrates, because of their relatively constant body temperature, and he suggests examining related physiological information in this group if we want to understand the size-dependency of metabolic rate. In searching for constraints on power use we shall consider the two extremes of inactive animals, and animals using maximum power moving as fast as they can.

If inactive animals only used metabolism to maintain body temperature and heat was only lost by radiation then metabolic rate would scale as the surface area of the skin, since heat loss per unit skin area is constant, i.e. (body size)$^{0.67}$. This places a lower constraint on the size-dependency of metabolic rate. A complication is that heat loss might also occur by conduction and convection, but these processes are likely to lead to downward deviation from 0.67 (Schmidt-Nielsen, 1984).

At the other extreme, animals use most power when moving as fast as they can, and this power goes to fuel the muscles (cf. Chapter 3). Power is rate of doing work, and work is force times distance. We therefore need to know the force which muscles can exert, and how far and how often they contract, i.e. contraction distance and frequency. The microscopic design of muscle places limits on the maximum force it can exert and on contraction distance. Evolution is likely to have maximized both of these microscopic properties of muscle, since they directly affect speed of movement, and there is a big pay-off from being able to move as fast as possible. It therefore comes as no surprise to find that they are body-size-independent (Schmidt-Nielsen, 1984), and we imagine these values represent constraints on evolution.

More specifically, the maximum force per unit cross section that can be exerted by any muscle depends on the microscopic structure of the muscle, and this is uniform in the higher vertebrates. Thus the number of filaments per cross-sectional area of muscle is the same, filament thickness is constant, sarcomere length, the length of the filaments, and the maximum overlap between thick and thin filaments, are all of the same magnitude in small and large vertebrates. As a result the maximum force which can be exerted per unit cross section is the same, and so is the maximum shortening per contraction.

It follows that maximum work (force × distance) performed in one contraction is scale-independent when calculated per unit volume of muscle. Calculations can now be made about the size-dependency of power in animals moving as fast as they can. Power = rate of doing work = (work per unit volume of muscle per contraction) × (volume of muscle) × (frequency of contractions). As we have seen, work per unit volume of muscle per contraction is a constant, i.e. size-independent. Volume of muscle is directly related to body size (scales as $m^{1.0}$) (Alexander *et al.*, 1981), and frequency of contractions is the same as stride frequency, which scales as $m^{-0.18}$ in galloping antelope and running birds, (Maloiy *et al.*, 1979).

Note that we have no explanation for the scale-dependency of stride frequency, we have to accept it as a 'given' derived from observations. It follows that maximum power should scale as $m^{1.0-0.18}$ = $m^{0.82}$. For comparison the actual value appears to be very close, $m^{0.845\pm0.015}$ (Taylor *et al.*, 1981).

We have identified two constraints on the size-dependency of metabolic rate; a lower constraint of $m^{0.67}$ (or possibly lower) for heat loss, and an upper constraint of $m^{0.82}$ for animals moving as fast as they can. Hence average metabolic rate should scale between $m^{0.67}$ and $m^{0.82}$, which it does (Fig. 6.2).

Most groups of animals, however, do not lie unambiguously on either constraint, and $m^{0.75}$ seems to be the most common relationship (Fig. 6.2). Another interesting and related observation is that physiological times — e.g. life times, gestation times and heart beat duration, — scale with $m^{0.25}$, which is body mass divided by the scaling dependency of metabolic rate ($m^{1-0.75}$). These findings provide some grounds for thinking that optimization may be more important than constraint

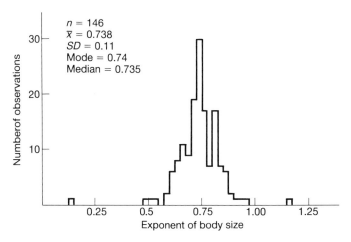

Fig. 6.2. Frequency distribution of the exponents of body mass in allometric relations for metabolic rate. (After Peters, 1983).

in the evolution of metabolic rate (i.e. type ii deviations). Adaptive explanations of the 0.75 rule are reviewed in Peters (1983) and Schmidt-Nielsen (1984), and Calder (1984) and McMahon & Bonner (1983) both provide their own explanations. For the rest of this chapter we shall assume that average and basal metabolic rate (BMR) scale as $m^{0.75}$.

6.2.2 CONSTRAINTS ON MUSCLE FORM AND FUNCTION

In the last section we noted that the maximum force which can be developed in muscle is the same for large and small vertebrates. This has implications for muscle morphology in animals which walk or run, and these are explored below. The arguments are based on those of Alexander and his co-workers.

When an animal is running at maximum speed the force per unit cross section developed in its leg muscles depends on
— the proportion of the stride when the limb is in contact with the ground (the so-called duty factor, β)
— the cross-sectional area of the muscles
— the animal's body mass, m
— the nature of the ground.

The forces on the feet of running animals are proportional to m/β; these are greatest at maximum speed, and then the duty factor, β, is at a minimum. The only available data on duty factors are from Alexander *et al.* 1981 and suggest that at maximum speed they scale as $m^{0.11}$ (fore feet) and $m^{0.14}$ (hind feet). These data should be treated with some caution, however, because they relate to ungulates, particularly Bovidae, whose bones are known to scale differently from those of other mammals (Alexander *et al.*, 1979).

If duty factors do scale as $m^{0.1}$ — and this we have to take as 'given' because there is no *a priori* argument at present — then the forces on the feet of running animals must scale as $m^{1-0.1} = m^{0.9}$. Since the lengths of the leg bones of mammals and running birds scale as $m^{0.35}$ (Alexander *et al.*, 1979; Maloiy *et al.*, 1979) — again a 'given' derived from observations (Fig. 6.3) — then the moment about the joint exerted by forces on the feet must scale as $m^{0.9} \times m^{0.35} = m^{1.25}$ or possibly slightly less, $m^{1.2}$, after allowing for the fact that large mammals tend to run on straighter legs than smaller mammals (Alexander *et al.*, 1981). This moment is counterbalanced by the force in the leg muscles acting on a bone lever (whose length is called by engineers a 'moment arm'; see Fig. 6.4). The minimum force is achieved

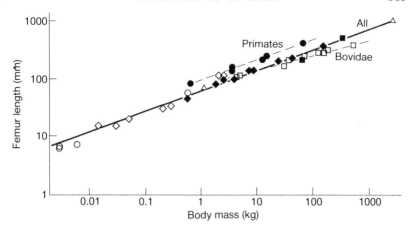

Fig. 6.3.   Log plot of femur length against body mass for various mammal groups. ○, Insectivora; ●, Primates; ◇, Rodentia; ◆, Fissipedia; □, Bovidae; ■, other Artiodactyla; △, other orders (Lagomorpha and Proboscidea). The solid line has slope 0.36 ± 0.06. After Alexander *et al.* (1979).

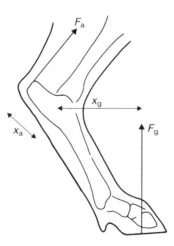

Fig. 6.4.   Maximum moments about the ankle joint in a buffalo galloping. F_a = force in the ankle extensor muscle, x_a = distance from the joint at which this force acts, i.e. the length of the 'bone lever' referred to in the text. The moment exerted by the ankle extensor muscle is $F_a x_a$. F_g = force on the foot, x_g = distance from the joint at which this force acts. The moment about the ankle exerted by the force on the foot is $F_g x_g$. Since the two moments have to be equal, $F_a x_a = F_g x_g$. This leads to the conclusion that limb radius, r, scales as $m^{0.4}$, as shown in the text. After Alexander *et al.* (1979).

116 CHAPTER 6

when the bone lever is as long as possible — i.e. roughly the radius of the limb, which we will call r. Since the cross-sectional area of the muscle is proportional to r^2, the maximum force exerted by the muscle is also proportional to r^2. Hence the moment achieved by the leg muscles is proportional to r^3, and this counteracts the moment about the joint exerted by the force on the feet which, as shown above, scales as $m^{1.2}$. Hence r^3 should scale as $m^{1.2}$, and therefore r should scale as $m^{0.4}$, and the cross-sectional area of leg muscle, which scales as r^2, should scale as $m^{0.8}$ — both predictions being true in mammals (Alexander *et al.*, 1981) and running birds (examples in Fig. 6.5).

6.2.3 CONSTRAINTS ON FOOD INPUT AND PRODUCTION

Feeding rate may be subject to a limit which varies with body mass m. Since absorption is a surface phenomenon, limited by the transfer of nutrients across the gut, it could be predicted that food intake should scale as $m^{0.67}$, but since digestive enzymes operate on the whole volume of food in the gut it could also be argued that food intake should scale linearly with m. It has usually been assumed that on an interspecific basis feeding rate scales between these 2 limits, i.e. roughly as $m^{0.75}$ like metabolic rate. For example Drent and Daan (1980) cautiously suggest a ceiling of 4 × basal metabolic rate on the rate at which energy can be acquired (based on data from 5 species of bird and from man). Actual feeding rates do scale as $m^{0.75}$ (Table 6.1). Note, how-

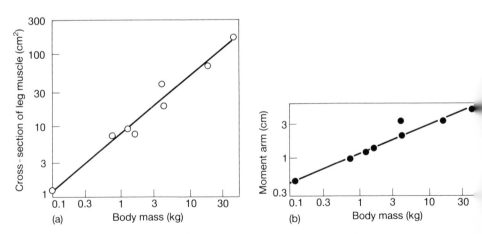

Fig. 6.5. Log plot against body mass of (a) the cross-sectional (fibre) area of the gastrocnemius muscle and (b) its moment arm about the ankle. The slopes are 0.81 ± 0.14 and 0.36 ± 0.11. After Maloiy *et al.* (1979).

Table 6.1. Daily food intake of mammals and birds in captivity (abbreviated from Calder, 1984)

Group	Food intake
Mammals	
Vegetable and seed eating	kg food/d $= 0.157\ m^{0.84}$
Animal-food eating	kg food/d $= 0.234\ m^{0.72}$
Herbivores	kJ/d $= 971\ m^{0.73}$
Carnivores	kJ/d $= 975\ m^{0.70}$
Mean	$\propto\ m^{0.75}$
Birds and mammals	
Herbivores	kJ/d $= 1006\ m^{0.72}$
Carnivores	kJ/d $=\ 917\ m^{0.69}$
Mean	$\propto\ m^{0.71}$
Maximum	kJ/d $= 1713\ m^{0.72}$

ever, that on an intraspecific basis, food input cannot scale perfectly with metabolic rate, otherwise all growth curves would be unlimited; here scaling is usually $m^{<0.75}$ (Chapter 5). Fig. 6.6 illustrates how scaling to $m^{0.75}$ on an interspecific basis is compatible with scaling to $m^{<0.75}$ on an intraspecific basis. In practice, intraspecific allometric regressions often have shallower slopes than interspecific ones (Gould, 1966; Leutenegger, 1976; Clutton-Brock & Harvey, 1979).

Nevertheless, feeding rate is subject to a limit which scales as $m^{0.75}$ interspecifically, and if maintenance metabolic rate also scales as $m^{0.75}$, then the resources available for production (i.e. growth and reproduction) must also be subject to a limit which scales as $m^{0.75}$ interspecifically. Peters (1983 p. 124) reckons actual growth rates scale as $m^{0.72}$ in mammals (Table 6.2), and this might mean that growth rates are maximized to their physiological limits in this group. However, there are type ii downward deviations from the slope of 0.72, particularly in apes, ruminants and ungulates, and similar deviations occur in reptiles and fishes (Case, 1978) and possibly invertebrates (Calow & Townsend, 1981b). This might mean that the mortality risks associated with rapid growth (Chapter 5) become more important in bigger animals and/or that a bigger starting size means that risks to extrinsic mortality factors are lessened so that the intrinsic risks associated with rapid growth are less worth taking.

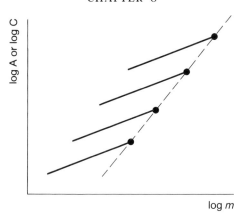

Fig. 6.6. Relationship between intraspecific (———) and interspecific (– –) allometries of food input against body mass. The interspecific relationships are often based on data from adults (●), but they could link other life-cycle stages with the same result.

Table 6.2. Allometric relationship between growth rate and adult body mass. Growth rate is determined as the average growth rate as an animal grows from 10 to 90% of its adult body mass. Abbreviated from Peters (1983)

Taxon	N	Slope	Proportion of variation explained
Animals		0.74	
Mammals	167	0.72	0.92
Eutheria	162	0.72	0.92
Marsupials	4	0.82	0.97
Insectivores	8	0.67	0.76
Primates	16	0.37	0.82
Prosimians	5	0.62	0.86
Anthropoids	11	0.35	0.79
Chiroptera	10	0.65	0.64
Rodents	60	0.67	0.88
Hystricomorphs	9	0.67	0.94
Myomorphs	32	0.72	0.84
Sciuromorphs	17	0.82	0.70
Lagomorphs	9	0.61	0.82
Fissipeds	23	0.70	0.86
Pinnipeds	11	0.74	0.38
Cetaceans	3	0.75	0.99
Ruminants	17	0.43	0.62
Ungulates	20	0.52	0.69
Birds	50	0.722	0.89
Reptiles	43	0.67	0.86
Fish	10	0.61	0.62

Reproductive growth (defined as litter mass/gestation time) should also be subject to a limit scaling as $m^{0.75}$ (Peters, 1983; Reiss, 1985). However actual reproductive growth in mammals scales as $m^{0.55}$ (Peters, 1983 pp 124−125). Similarly, litter mass (which has to be carried around by the adult) at birth might be constrained below some limiting fraction of body mass, so the constraint should scale as $m^{1.00}$, yet actual litter masses at birth scale at less than this in many animal groups (Table 6.3). These downward deviations from the expected slopes again imply that, in those groups where they apply, the investments are not maximized to physiological limits in larger animals (type ii deviations). This suggests either that the costs of reproduction are more serious for larger animals and/or that they are less worth paying and this might be why iteroparity, rather than semelparity, is a more prominent feature of larger animals.

Table 6.3. Allometric relationship between energy invested in reproduction (P_r) and body mass. Note that all of the exponents are less than one (see text). N is the number of species used in each regression. (Abbreviated from Reiss, 1985)

Taxon	Measurement of P_r	Exponent	Correlation coefficient	Weight range	N
Spiders	Clutch number	0.84	0.94		28
Hoverflies	Clutch volume	0.95	0.89		30
Poikilotherms	Clutch volume	0.92	0.93	3 mg−140 kg	
Salamanders	Clutch volume	0.64	0.90	140 mg−8900 g	74
Frogs	Clutch volume	0.90	0.98	0.7−120 g	23
Reptiles	Litter weight	0.88	0.98		
Birds, Anatidae	Clutch weight	0.52	0.81	310 g − 12 kg	149
Birds, Phasianidae	Clutch weight	0.53	0.80	45 g−5 kg	50
Birds	Clutch weight	0.74	0.92	4 g−100 kg	

6.2.4 EFFICIENCY IN RELATION TO BODY SIZE

One might have thought that, in performing their various functions, larger organs would be at least as efficient as smaller ones, when compared per unit organ weight. This might be expected on the grounds that if a single large organ were relatively inefficient it would be replaced by a number of smaller organs (this is part of the reason why big ships have a number of relatively small engines; McMahon & Bonner, 1983). However this general prediction appears to be false; lung and circulatory system volumes, for example, scale as $m^{1.00}$ in mammals (Fig. 6.7) and birds (Calder, 1984), but power supplied to

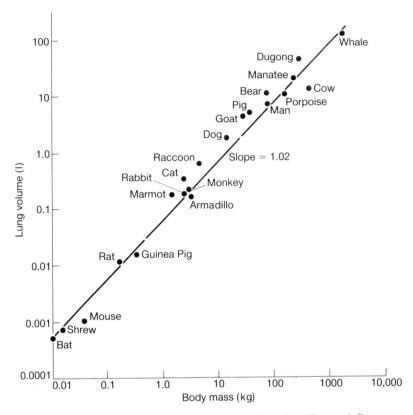

Fig. 6.7. Log plot of lung volume against body mass. Data from Tenney & Remmers (1963).

the body scales as $m^{0.75}$, so specific performances (power supply/weight carried) decline as $m^{-0.25}$. This is probably because larger animals are forced to adopt the same body plan as smaller animals, with the result that breathing rate necessarily declines, specific rate of turn-over of oxygen in the lungs declines, and so the partial pressure of oxygen in the lungs declines with body size (Peters, 1983 pp 48–51; Schmidt-Nielsen, 1984). Arguments of this type may also apply to the circulatory system.

In similar fashion the power supplied to the body by the digestive system scales as $m^{0.75}$, but gut volume scales as $m^{1.0}$ or m^{1+} in most mammal groups (Fig. 6.8). Thus specific performance (power supply/weight carried) declines as $m^{-0.25}$. This might be an example of a general decline in performance with organ size, as with the lung and

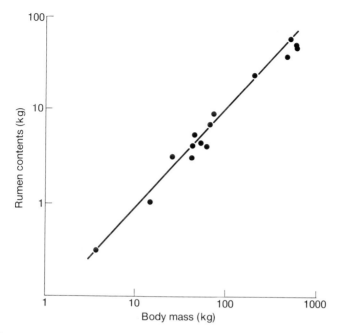

Fig. 6.8. Log plot of rumen contents against body mass for different species of ruminants. Slope = 1.08. After Clutton-Brock & Harvey (1983).

circulatory systems described above, but it might also result from bigger animals being forced to take worse food than smaller ones. For example, small animals need less food and can therefore take the time necessary to select only better quality food items. They thus obtain a high quality diet, and on the basis of the arguments of Chapter 2 would be expected to have relatively small guts. Large animals 'cannot afford to be so choosey' and are therefore more likely to take worse food. In this situation their optimal strategy is to have relatively large guts (Chapter 2). On the basis of this theory it should, of course, be possible to demonstrate that larger animals eat worse food. This seems to be the case generally for the herbivorous, frugivorous and omnivorous mammals (Clutton-Brock & Harvey, 1983; cf. Demment & Soest, 1985).

Thus declining food quality may contribute to the decline in gut performance with body size. It seems unlikely to be the whole story, however, since an exactly analogous decline in performance was found in the lung and circulatory system where the resource input to the system — air — is obviously the same for all air-breathing animals irrespective of size. Thus, here performance is known to decline with organ size even though resource quality is known not to vary.

6.2.5 CONCLUSIONS AND SUMMARY

In conclusion, then, the interspecific comparisons discussed above can suggest some type ii deviations from the null hypotheses. The form of these deviations suggests further hypotheses about the nature of the costs associated, for example, with rapid growth and high levels of investment in reproduction and these can, at least in principle, be considered further by the more detailed programmes described in Chapter 4 (for reproduction) and 5 (for growth). Deviations from *a priori* expectations about organ allometries might also be due to optimization phenomena (as in the gut) and these again suggested predictions (e.g. that larger animals should eat poorer quality food than smaller animals) that can be tested. Before it can be taken too seriously, however, the programme needs sharpening up in two main respects. First, the *a priori* arguments, that establish the null hypotheses, require to be made physically and biologically convincing. This is a challenge for physiology. Second, the comparisons between null hypothesis and observations need to be made statistically convincing. We have skated over this aspect intentionally — because it is not straightfoward and probably needs further refining.

6.3   On when to stop growing — the optimal body size

6.3.1 THE MODEL

An increase in size potentially has two opposing effects on fitness. It usually requires longer development (negative effect) but might yield increased fecundity or decreased adult mortality rate (positive effect). Increased size might also be achieved by increasing growth rate (and thus mortality risk) but this possibility was considered in Chapters 4 and 5. An example of a size–fecundity relationship in a crab is shown in Fig. 6.9.

Some possible relationships between size and developmental time, and size and fecundity are illustrated in Fig. 6.10 with the ones that we consider to be most plausible marked A. Curve A in Fig. 6.10a is purely speculative and its shape will definitely depend on the mating system, the extent of fighting between males for access to females, etc. Curve A in Fig. 6.10b suggests that the final increments in size take a

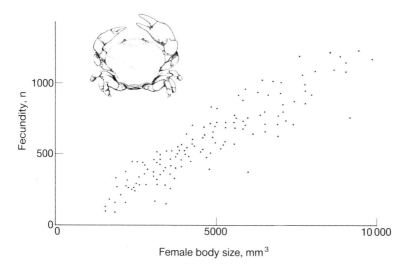

Fig. 6.9. Relationship between fecundity and female body size, measured as (carapace width)^3, in a brachyuran crab. After Jones (1978), see also Sastry (1983).

disproportionate length of time, as in the final phase of a sigmoid growth curve (see Chapter 5). The important trade-off is between fecundity, n, and developmental time t_j, with size acting as a hidden intermediary (Fig. 6.10(c)). Contours of equal F can be calculated from the formula:

$$t_j = \frac{1}{F + \mu_j}(\log_e n - \log_e(1 - e^{-(F+\mu_a)t_a})) \qquad (6.1)$$

where μ_j = juvenile mortality rate, μ_a = adult mortality rate, n = number of offspring produced per breeding attempt, and t_a is the interval between breeding attempts. (Equation 6.1 is derived from Equation 4.2). F-contours are shown in Fig. 6.10d for hypothetical values of μ_a, μ_j and t_a.

From Fig. 6.10 it can be seen that to obtain equivalent fitness, fecundity has to increase increasingly with developmental time. Specifically, n has to increase exponentially with t_j, $n \propto \exp(t_j)$, as can be seen by rearranging equation 6.1. This result was shown for a particular case in a seminal paper by Lewontin (1965) on colonizing species but its generality has been questioned on a number of grounds. In particular:

1 Development time becomes irrelevant in seasonal situations where the time between breeding seasons is fixed (Meats, 1971). Nevertheless,

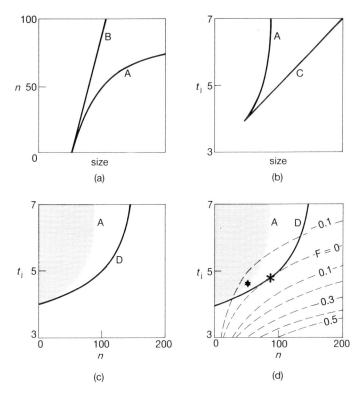

Fig. 6.10. Possible relationships between (a) size and fecundity, (b) size and development time. The relationships we consider most likely are marked A (see text). Eliminating size between (a) and (b) gives the $n-t_j$ trade-off (c). In (d) the broken lines are lines of equal F calculated from equation (6.1) with $\mu_j = 0.921$, $\mu_a = 0.693$ and $t_a = 1$. See text for further details.

even under these circumstances we would expect: (a) the time between birth and breeding to be reduced to the time between breeding seasons and (b) for there to be an advantage in starting to breed as early as possible in the breeding season.

2 Shortening development is unimportant in non-expanding populations and would be disadvantageous in contracting populations. The rationale behind the latter is that in a contracting population it pays to delay breeding because there are then fewer members of the population and so offspring of delayed breeders represent a larger fraction of the total population and hence, by definition, are fitter. This is at the heart of arguments made by Hamilton (1966), Charlesworth (1980) and others. The flaw in it, however, is that it assumes that pre-reproductive mortality is *fixed*. In our view it is much more plausible that adult mortality rate per unit time is constant, and this is supported by the life-table data (Itô, 1980, Fig. 5.8). As soon as this is brought into the model it pays to shorten the time between birth and breeding because this reduces the chances of dying before breeding. Thus, the longer the time over which a constant mortality rate operates, the more die.

Now we try to define those conditions that will favour an increase in final, adult size. This can be achieved if either (a) the relationship between developmental time and size becomes more nearly linear (curve C in Fig. 6.10b) and/or (b) the relationship between fecundity and size becomes more nearly linear (curve B in Fig. 6.10a). Either effect enlarges the options set in the $n-t_j$ plane and changes the trade-off curve towards curve D in Fig. 6.10c.

Fig. 6.10d shows that the optimal strategy is then a big increase in fecundity, and negligible or small increase in the optimal value of t_j. The implications for body size depend on what caused the enlargement of the options set from A to D. If the improvement was brought about by an improvement in the size$-t_j$ relationship (curve A → C in Fig. 6.10b) then larger size is selected for (because optimal t_j is nearly the same for curves A and D in Fig. 6.10d. Referring back to Fig. 6.10b, size at a given value of t_j is larger on curve C than on A). On the other hand if the improvement A → D was brought about by an improvement in the size$-n$ relationship (curve A → B in Fig. 6.10a) then optimal size increases only slightly if at all (because optimal t_j is only very slightly longer for curve D than for A. Size is given by curve A in Fig. 6.10b and a slight increase in t_j produces a slight increase in body size).

An example of this sort of analysis is provided in a pioneering case study of optimal body size in *Drosophila melanogaster* by Roff (1981). Roff made specific assumptions about the relationships between body size and fecundity, and body size and development time, and was therefore able to discover the relationship between body size and fitness, which is depicted in Fig. 6.11. The inclusion of the optimal size of 0.95 mm in the observed range from 0.90 mm to 1.15 mm is encouraging, although sensitivity analysis shows that the model is quite sensitive to one of its parameters (Roff, 1981; Ricklefs, 1982).

6.3.2 CONCLUSIONS AND SUMMARY

Optimal size depends upon the relationship between size and developmental time, and size and fecundity. As usual we have to rely more on educated guesses for the form of these relationships than on real data.

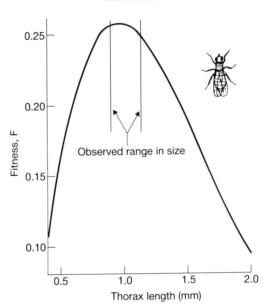

Fig. 6.11. Optimal body size in the fruit fly *Drosophila melanogaster*. See text for discussion. After Roff (1981).

Nevertheless, the model clearly illustrates that for the evolution of larger size there has to be an appreciable increase in fecundity for a small increase in developmental time. Possible scenarios are that mutations appear that have the necessary enabling effects on morphology and development and/or there is a shift in niche that allows improved fecundities at larger size.

7
Classifying Habitats by Selection Pressure

7.1 Introduction

Classically, physiological ecologists have paid much attention to classifying habitats by selection pressures; i.e. explaining the characteristics of organisms according to the dominant forces of selection operating in particular ecological circumstances. This is the *a posteriori* approach (Chapter 1; see particularly Table 1.1). We have now developed an extensive theoretical framework from which it is possible to make specific predictions about the kinds of traits that should be expected in particular ecological circumstances. This is the *a priori* approach. Here, we attempt to generalize these predictions into a classification of habitats based upon the anticipated effects of a few, key selection pressures. In so doing, we follow the lead of Robert MacArthur (see particularly his 1972 book). He distinguished between habitats according to whether they were associated with '*r*' or '*K*' selection. Since this classification has been enormously influential, we first provide a short explanation of it and later indicate how it relates to our own classification.

7.2 *r* and *K* selection

In its most basic form, MacArthur's theory is concerned with the effects of population density on general demographic traits, especially on fitness, *F*, measured by rate of increase per individual. The simplest possible relation between this measure of fitness and density is the straight line:

(rate of increase per individual) = $r - r/K$ (density),

where *r* and *K* are parameters characterized below. Writing density = *N* gives

$$\frac{1}{N}\frac{dN}{dt} = r - \frac{r}{K}N \qquad (7.1)$$

127

$$\therefore \frac{dN}{dt} = N\left(r - \frac{rN}{K}\right) \qquad (7.2)$$

which provides the starting point for contemporary population dynamics. Rate of increase per individual is the measure of fitness (F) that we have used throughout. Here, then, we have another point of contact between population dynamics and population genetics (see also Chapter 1).

Characters that bring about high F at low density might be the same as those bringing about high F at high density and if this were so there would be no more to say. On the other hand, if superior fitness at low density is incompatible with superior fitness at high density, then two distinct types of organisms (A and B) may evolve; A being fitter than B at low density and B being fitter than A at high density. In Fig. 7.1, rate of increase per individual at zero density is labelled r, and density when rate of increase per individual is zero is labelled K (carrying capacity). A increases at a faster rate at low density ($r_A > r_B$) and is said to be r-selected. On the other hand, B increases at a faster rate at high density provided its carrying capacity is higher ($K_B > K_A$) (MacArthur, 1972; Sibly & Calow, 1983), and is said to be K-selected.

However, a classification of selection pressures based only on the effects of density on fitness cannot be very precise because the latter is made up of a number of components (fecundity, survivorship, growth rate of offspring), different values of which might maximize fitness at

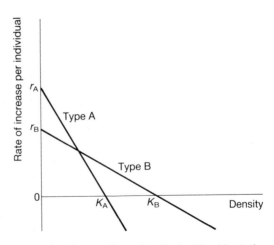

Fig. 7.1. r- and K-selection (see text for explanation). After MacArthur, 1972.

either low or high density. Moreover, the density effects themselves might occur in different ways (e.g. as a result of competition, predation or dispersal) and the consequences of one might not be the same as those of another (Boyce, 1984). Only by the addition of a number of auxiliary hypotheses has it been possible to construct classifications rich with predictions about phenotypic traits. Thus, the discussions in Chapters 1 and 3 indicate that organisms have to allocate resources between competing metabolic demands and this leads to a trade-off between fecundity-promoting and survival-promoting traits. The suggestion of an incompatibility between high rates and high efficiencies of production also predicts precise distinctions between traits that should evolve under conditions with and without resource limitations (Chapter 3). Pianka (1970) seems to have made a number of implicit assumptions of this kind in constructing his detailed catalogue of differences between the consequences of r and K selection (Table 7.1).

Recognizing the limitations of a classification based only on population density, other authors have tried to enrich the theory by incorporating other forces. Whittaker (1975), for example, pointed out that some populations occur in harsh environments and that environmental stress might have been a dominant force in shaping their evolution. This idea is also incorporated into the classificatory schemes of Grime (1974, 1979), Southwood (1977) and Greenslade (1983).

7.3   Classification based on theory in Chapter 4

The problem with many of the classifications noted above is that they are not defined very precisely and many are probably *a posteriori* rather than *a priori*. This section attempts to develop a more rigorous classification, *a priori*, from principles introduced in earlier chapters. We shall conclude that a one-dimensional classification, by population density, should be replaced by a two-dimensional scheme. Two key variables will be identified (1) an index of age-specific survivorship (S) and (2) an index of growth rate of offspring (G). Both S and G are determined by environmental conditions (extrinsic factors) but are also influenced by internal constraints and trade-offs (intrinsic factors). By definition, habitat classification is more concerned with extrinsic factors, but in practice extrinsic and intrinsic factors are often difficult to distinguish. Therefore, after defining S and G in terms of the theory, we offer operational definitions as the first stage towards applying the classification in the real world.

As in Chapter 4, we concentrate on absorption costing and will

Table 7.1. Some correlates of r- and K-selection. From Pianka (1970)

	r-Selection	K-Selection
Climate	Variable and/or unpredictable: uncertain	Fairly constant and/or predictable: more certain
Mortality	Often catastrophic, nondirected, density-independent	More directed, density-dependent
Survivorship	Often Type III (Deevey, 1947)	Usually Type I and II (Deevey, 1947)
Population size	Variable in time, nonequilibrium; usually well below carrying capacity of environment; unsaturated communities or portions thereof; ecologic vacuums; recolonization each year	Fairly constant in time, equilibrium; at or near carrying capacity of the environment; saturated communities; no recolonization necessary
Intra- and interspecific competition	Variable, often lax	Usually keen
Relative abundance	Often does not fit MacArthur's broken stick model	Frequently fits the MacArthur model
Selection favours	1 Rapid development 2 High r_{max} 3 Early reproduction 4 Small body size 5 Semelparity: single reproduction	1 Slower development, greater competitive ability 2 Lower resource thresholds 3 Delayed reproduction 4 Larger body size 5 Iteroparity: repeated reproductions
Length of life	Short, usually less than 1 year	Longer, usually more than 1 year
Leads to	Productivity	Efficiency

discuss possible complications that arise from direct costing only in passing. In Chapter 4 we showed, using selective landscapes, that investment in reproduction should:

1 Increase with juvenile survivorship assuming adult survivorship (S_a) is constant (p. 71).

2 Reduce with adult survivorship assuming juvenile survivorship (S_j) is constant (p. 77).

From 1 and 2 it is clear that the investment in reproduction should increase as the ratio S_j/S_a increases. The rigorous argument (Sibly & Calow, 1985) in fact indicates that the investment should increase with

$$S = \frac{S_j}{S_a t_a} e^{-Ft_j + Ft_a} \tag{7.3}$$

but if $t_j = t_a = 1$ or F is small, both of which are plausible, the complex term reduces to the simpler ratio — and we refer to this index of age-specific survivorship as S. Here, far from complicating the argument, the Principle of Compensation can complement it (p. 22).

Also in Chapter 4, we showed that propagule size should increase as the environmental conditions for growth deteriorate. We refer to these conditions by another index, G, which describes potential growth rate (see equation (7.5) below). This is a measure of the extent to which the environment influences t_j, the time between birth and breeding.

Of these two predictions, only the one involving a trade-off between S_a and n is influenced by direct costing (Chapter 4). Under these conditions it can be shown (Sibly & Calow, 1985) that S is calculated from

$$S = \frac{S_j S_a}{t_a} e^{-Ft_j - Ft_a} \tag{7.4}$$

That is, as already made clear, high adult survivorship in these circumstances favours a higher investment in reproduction rather than low adult survivorship. Note, though, that the Principle of Compensation can cause complications for this model (p. 22).

7.4 Derivation of S and G under absorption costing

The index S is derived from information on life tables which can be obtained from routine sampling programmes (Southwood, 1978).

S_j is the probability of individuals surviving between birth and their own first breeding.

S_a is the overall adult survivorship, between breeding seasons. It is made up of intrinsic (S_{in}) and extrinsic components (S_{ex}).

Assuming, that the two are independent then, as already noted (p. 76), $S_a = S_{in} \cdot S_{ex}$. Since, for a given physiology, the relationship between S_{in} and n is fixed by the trade-off curve, then it can be considered a constant, and so S_a depends on S_{ex}, i.e. it changes with extrinsic survival probability between habitats. Hence we need to measure changes in extrinsic mortality levels. Clearly, intrinsic mortality must be excluded because, for semelparous organisms, S_{in} will by definition be zero. In this case S_a is zero, and this automatically predicts semelparity under absorption costing — but the argument is clearly circular. The correct argument is that S_{in} is low only because high n is favoured, and this is favoured only because S_{ex} is low as compared with S_j. It will, of course, be difficult to separate S_{ex} from S_{in}! However, there are two possible ways of doing it:

(a)   to quantify S_{ex} from information on mortality agents. Parry (1982) did this for limpets and found a reasonable correlation between S_{ex} so estimated, and investment in reproduction (Fig. 7.2).

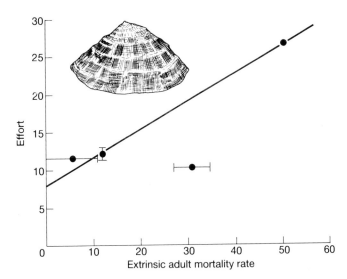

Fig. 7.2.   Reproductive effort (measured as energy invested in gametes ÷ energy absorbed by parents x 100) against extrinsic mortality for four species of limpet in S.E. Australia. Extrinsic adult mortality rates were inferred from total adult mortality rates, a knowledge of the causes of mortality, and the likely effects that a change in reproductive allocation would have on the vulnerability of each species to its observed agents of mortality. Each point represents a separate species. There are grounds for thinking the bottom right point should be further to the left, which would improve the correlation. Error bars are ranges of values. Redrawn from Parry (1982).

(b)   by measuring S_a for adult-like (i.e. same sized) animals, but at a part in the life cycle when there are unlikely to be significant reproductive influences. Maltby used this technique for leeches, extrapolating mortality rates in pre-breeding adults to between-breeding periods. She found a reasonable correlation between S values calculated on this basis and degree of iteroparity in populations (Table 7.2) but there are complications because the trade-off curves in these cases turn out to be linear rather than convex (Maltby & Calow, in preparation).

Table 7.2.   Observations on 2 populations of one species of freshwater leech (*Erpobdella octoculata*) in rivers in the West of Scotland (Maltby & Calow, 1986)

Population	$S^{\dagger}$	'Investment in reproduction' (young parent^{-1} breeding season^{-1})	Mode of reproduction
Allander water	0.098	24.6	Iteroparous
Blane water	0.389	44.2	Semelparous

† Defined in text

G is obtained from observations on individual growth rates in natural populations:

$$G = \frac{\log_e m(t_y) - \log_e m(t_x)}{t_y - t_x} \tag{7.5}$$

Where t_x = age at which offspring gain independence, t_y = some slightly later age, and $m(t)$ = mass of offspring at age t. An index of this kind is commonly used in assessing growth curves (Hunt, 1982). Note that both S and G are indices which summarize selection pressures exerted in particular habitats. However, note also that they are assessed from measurements on organisms experiencing the selection pressures. Hence the organisms are being used as measuring devices for assessing the strength of conditions influencing them. Because these influences not only have proximate effects (which are the ones we are interested in measuring) but also ultimate effects (i.e. genetic changes, which are the ones we are interested in predicting) some care has to be exercised in interpreting these data. Attention has already been drawn to this, particularly for S_a.

7.5   Habitat classification based on S and G

The focus is, therefore, shifting away from population density as a single classifying variable towards two key parameters which directly affect life-cycle strategy; in particular S and G, as defined above. This suggests a two- rather than one-dimensional classification and a simple matrix can be generated by dividing the two axes, G and S, into high and low levels (Table 7.3). It is now possible to distinguish between four categories of selection: high G, high S; low G, low S; high G, low S; low G, high S. At high S the total investment of resources in reproduction should be high and vice versa at low S. Similarly, at high G propagule size should be small and vice versa at low G. Combining these two predictions, it follows that fecundity (n) should be highest at high G, high S and lowest at low G, low S, with other habitat types favouring intermediate values.

Strictly, the two-dimensional $G-S$ continuum should be discussed as a whole, but it is easier to consider pairwise comparisons and to compare them with the one-dimensional r/K classification. The possible comparisons are therefore: low S and G vs. high S and G; low S and G

Table 7.3. Life-cycle predictions in relation to a dichotomous classification of habitats by G, an index of growth rate of offspring, and S, an index of survivorship (see text).

		P_r low	P_r high
High		z low	z low
		n intermediate	n very high
G			
Low		P_r low	P_r high
		z high	z high
		n low	n intermediate
		Low	High
			S

P_r = total investment in reproduction during a particular breeding attempt;
z = investment per egg; n = number of eggs.

vs. high S and low G; low S and high G vs. high S and G; low S and G vs. low S and high G; high S and low G vs. high S and G; high S and low G vs. low S and high G. Complications might arise, however, because resource availability can influence the investment in reproduction as well as G. Higher resource availability might mean that more offspring could be obtained at lower cost to the adult, e.g. if there were less drain on adult body reserves. This would shift the trade-off curve, $S_a(n)$, relating offspring number to adult survivorship, and the optimum is likely to involve the production of more offspring. If in this case higher resource availability for adults was associated with good growth conditions for offspring, then the effect would be to increase reproductive effort, P_r, and fecundity, n, on the top row of Table 7.3. Complications for predictions about P_r, due to resource availability, may therefore apply where there are gradients in G. When this happens at low S it is likely that P_r will not remain constant but will be very low at low G, low S. Similarly, at high S, high G is likely to allow a very high P_r and low G a lower P_r. These effects accentuate the expected differences in comparisons involving low G, low S vs. high G, high S, but blur expected differences in comparisons involving low G high S vs. high G low S. In this last instance it is difficult to know *a priori* whether differences in P_r will occur. Comparisons along this axis must therefore be treated with caution. Another possible effect of increased resource availability for adults is a decrease in time between breeding, t_j, which also favours the production of more offspring under some conditions.

7.6   Testing predictions

We here consider expectations at each level of S and G and indicate what evidence is available to evaluate them.

(1)   Low S, low G vs. high S, high G.

This dichotomy is most nearly equivalent to the r/K distinction *as perceived by Pianka* (1970) and is referred to by Stearns (1977) as the 'Accepted Scheme'. According to the review by Stearns (1977), the predicted outcome (see Table 7.1) is far from universal, and Stearns took this as evidence against r/K selection. However, Pianka's r/K dichotomy is only one possible consequence of density effects and hence the finding that it does not occur is not decisive in refuting their importance (Boyce, 1984).

(2)   Low S, *low* G *vs. high* S, *low* G.

This represents a gradient in survivorship when conditions for growth
are poor. For example, British freshwater triclads all suffer resource
limitations (Reynoldson, 1983). Yet because of differences in the physiology of hatchlings, the juvenile survivorship of some species is better
than others (Calow & Woollhead, 1977; Woollhead & Calow, 1979).
Absorption costing probably predominates here. On this basis, triclads
conform well to the predictions since species in which juveniles have
good survivorship invest more in reproduction than those that have
poor juvenile survivorship (Calow & Woollhead, 1977; Woollhead &
Calow, 1979; Woollhead, 1983). This kind of gradient in selection
pressures is likely to be common in resource-limited top carnivores.

(3)   Low S, *high* G *vs. high* S, *high* G.

This represents a gradient in survivorship when conditions for growth
are good. For example, gyrodactylid ectoparasites (Monogenea) of
aquatic vertebrates probably experience good, reasonably stable sup-
plies of food from their hosts, yet different species occupy different
sites which differ in vulnerability. Thus, *Gyrodactylus gasterostei* from
the skin of three-spined sticklebacks has higher adult mortality and
fecundity than *Gyrodicotylus gallieni* which lives in the more protected
site of the mouth of *Xenopus*. Gill parasites probably have intermediate
levels of mortality and fecundity. (These data are from P. Harris, cited
in Calow, 1983c). High growth rates are probably typical of many
parasites and so differences in *S* are likely to play an important part in
their life-cycle evolution.

(4)   Low S, *low* G *vs. low* S, *high* G.

This represents a gradient in growth-promoting potential when conditions for survival are poor.

(5)   High S, *low* G *vs. high* S, *high* G.

This represents a gradient in growth-promoting potential when conditions for survival are good. Categories (4) and (5) have important
implications for reproductive strategies and are considered separately
and more extensively in the next section.

(6)   High S, *low* G *vs. low* S, *high* G.

Here predictions are opposite to the 'Accepted Scheme' (comparison
(1), above). It is possible that this distinction applies to endoparasites
(which have good conditions for growth, but poor chances of juvenile
survival) and free-living relatives (poorer conditions for growth, but
better for juvenile survival — Calow, 1979, 1983c).

However, this is when extremely rich trophic conditions (high G)
can have a complicating influence on the investment in reproduction
and its influence on post-reproductive survival. Thus the good trophic
conditions might pull the trade-off curve to the right as shown in Fig.
7.3 for, under these circumstances, a particular level of investment in n
will have less impact on S_a than if resources were more limiting. Then,
as the figure shows, even a lower S_j can favour an increased reproductive output.

This particular comparison is also possibly representative of a general shift in selection pressure with trophic position from the low
survival but good growth conditions for herbivores to the higher survival but poorer growth conditions for carnivores (see also Wilbur *et
al.*, 1974).

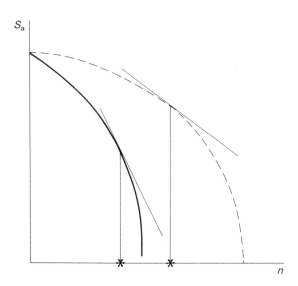

Fig. 7.3.   Trade-off between S_a and n of the same or similar systems under good
(– – –) and poor (——) trophic conditions.

7.7   Gradients of *G* and reproductive strategies

Gradients of *G* have important potential effects on the occurrence of different reproductive strategies. Since these are key features of life cycles we consider them separately here. We assume that *S* is held constant.

7.7.1 STRATEGIES OF NON-GAMETIC REPRODUCTION

By non-gametic reproduction we mean the mitotic production of re-productive propagules which are multicellular fragments of metazoans and unicellular products of protozoans. There are a variety of patterns ranging from simple division (fission) splitting into 2 (binary fission) or more (multicellular fission) propagules, budding, fragmentation, laceration and so on. The prediction is that fewer, larger fragments should be produced as conditions become poorer for individual growth. Hence, binary fission should be more common in these circumstances and multiple fission should appear as trophic circumstances improve.

Table 7.4 shows the distribution of binary and multiple fission between various taxa of parasitic and non-parasitic protozoans. We presume, as in the last section, that conditions for growth are better for parasites, because they live surrounded by food and there is, as can be seen, a highly significant association of multiple fission with this life-style. However, two reservations can be expressed about this conclusion. First, parasites might suffer poorer growth rates than free-living species because of immunological reactions of the host. However, the few data that we have on parasite growth rates do not support this conclusion (Calow, 1981). Second, there are taxonomic correlations in the data (only Sporozoa and Cnidosporidia are exclusively parasitic and carry out multiple fission, ciliate and mastigophoran parasites carry out binary

Table 7.4. Classification of protozoan species according to mode of reproduction and life-style, showing that multiple fission is associated with the parasitic life-style, and binary fission with the non-parasitic life-style. From Sibly & Calow (1982)

	Binary fission		Multiple fission	
	Ciliata	Mastigophora*	Sporozoa	Cnidosporidia
Parasites	1000	900	1100	900
Non-parasites	5000	2600	0	0

* A few Mastigophora reproduce by multiple fission.

fission) and this could mean that the results are more to do with taxonomic constraints than different selection pressures between parasitic and non-parasitic habitats. Nevertheless, the data are consistent with the model and strengthen our conviction about it. What should be attempted now is a more extensive analysis of the growth rates of parasites as con.pared with non-parasitic relatives, and a more extensive taxonomic analysis.

7.7.2 STRATEGIES OF GAMETE PRODUCTION

Gametes may be thought of as unicellular propagules produced by specific reproductive organs (gonads) either by meiotic or mitotic mechanisms: i.e. respectively sexual and parthenogenetic reproduction. We shall be concerned mainly with eggs and these can vary quite considerably in size depending upon how much nutritive reserves (usually albuminous yolk) they contain. The prediction is that a smaller number of larger eggs should be produced as conditions for post-hatching development deteriorate.

Kolding and Fenchel (1981) have found that winter breeding *Gammarus* produce smaller broods of larger eggs than summer breeding forms (Fig. 7.4), and crustaceans from colder waters (deeper, or nearer the poles) generally produce larger eggs than those from warmer waters (Sastry, 1983). Similarly, arctic and antarctic marine invertebrates tend to have fewer larger eggs than relatives living in temperate zones (Clark, 1979). In fishes, eggs incubated in colder months tend to be larger than those incubated in warmer months (Wootton, 1984). Finally parasitic tapeworms in which eggs are transferred to and develop in poikilothermic hosts (Fig. 7.5b and c) produce larger eggs than those in which the equivalent host is homeothermic (Fig. 7.5a). It is likely that homeothermic hosts provide better conditions than poikilothermic ones. Itô (1980) believes that correlations of this kind, all consistent with the predictions from the models, hold for most organisms.

7.7.3 GAMETE PRODUCTION VERSUS NON-GAMETIC
REPRODUCTION

The comparison here is between the production of non-gametic and gametic propagules as defined in 7.7.1 and 7.7.2 above. It is distinct, therefore, from the comparison between sexual and non-sexual (parthenogenetic) egg production, which can be carried out in terms of similar products (unicellular propagules), and which has stimulated

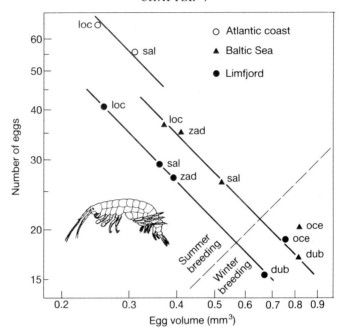

Fig. 7.4. Egg numbers (for females measuring 10 mm) and egg volumes for 5 species of *Gammarus* from three sampling areas. Winter breeders produced smaller broods of larger eggs than summer breeding forms, which probably enjoyed better conditions of juvenile growth (because of water temperature) and survival. dub = *G. duebeni*; loc = *G. locusta*; oce = *G. oceanicus*; sal = *G. salinus*; zad = *G. zaddachi*. From Kolding and Fenchel (1981).

much debate (Williams, 1975; Maynard Smith, 1978; Bell, 1982). In the following analysis, while recognizing its importance, we do not take into account the costs of sexuality.

In certain respects gamete production can be considered as an extreme form of fragmentation, involving the production of unicellular fragments. However, there are a number of complications. First, the gonads operate at a higher metabolic pace and efficiency than the somatic processes (either because respiration reduces relative to food input during breeding, or feeding increases relative to respiration, or because somatic stores and structures are used to supplement the food in forming the gametes or a combination of these (Calow, 1983d)). Second, and related to the first point, gamete production can cause reductions in the post-reproductive survival or the subsequent fecundity (n) of the parents (Chapter 4). Third, the gonads have to be built and maintained and yet the resources used in this process do not appear as

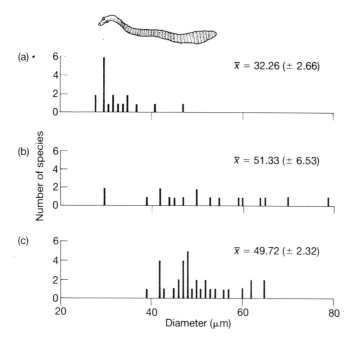

Fig. 7.5. Frequency distribution of egg size in cestode parasites, showing that those with homeothermic intermediate hosts (a) produce smaller eggs than those with poikilothermic intermediate hosts (b & c). a and b are cyclophyllideans; c are pseudophyllideans. From Calow (1983c) (data from Wardle & McLeod, 1952).

reproductive propagules. The production of gametes is rather like mass production, then, in that by investing in capital equipment (gonads), item cost can be reduced. In species in which gametes are produced, more reproductive propagules are likely to emerge per unit input of resources than would be possible by fragmentation and this now allows us to extend the model considerably. However, none of these complications affects the basic prediction that when gametic and non-gametic reproduction are possible within a particular taxonomic group, the former will be favoured if growing conditions are good and the latter if they are poor.

Freshwater flatworms are capable of both modes of reproduction. Table 7.5 shows that gamete production is more common than binary fission in eutrophic habitats and vice versa in oligotrophic habitats. Chia (1976) claims that non-gametic reproduction in anthozoan cnidarians becomes more common as the trophic conditions deteriorate.

When there is an alternation of gametic and non-gametic reproduc-

Table 7.5. Classification of the British and Japanese freshwater triclads showing that gamete production is associated with eutrophic habitats and binary fission with oligotrophic habitats ($\chi^2 = 6.5$ on combined data, $p < 0.02$). From Sibly & Calow (1982)

	British species			Japanese species	
	Oligotrophic (streams)	Eutrophic (lakes)		Oligotrophic (sub-terranean & cold-running streams	Eutrophic (lakes & slow-running streams)
Binary fission only	3	1	Binary fission sometimes	4	1
Gametes only	0	7	Gametes only	4	9

tion, as occurs in the life cycles of some organisms, we would expect smaller fragments to be produced when the curve of t_j against propagule size is shallower (Fig. 7.6). Very shallow curves would occur if growth is impossible over winter but very rapid in early spring. Under these conditions we would expect gamete production in the autumn or early winter. For the rest of the year, if growth is relatively slow the trade-off curve would be relatively steep and non-gametic reproduction would be expected from late spring to summer. Clearly all depends on the seasonal occurrence of good and bad growing conditions and, as yet, we have not found data from natural populations that are good enough to test these predictions.

The conventional explanation for the seasonal occurrence of gametic and non-gametic reproduction has been in terms of a posited seasonal advantage of sexual reproduction (references above), but the present analysis shows that other explanations are possible. Nevertheless our model will eventually need to be extended to take into account the genetic costs and benefits of sexual reproduction.

7.8 Conclusions and summary

Classifying habitats according to selection pressures has been a major goal of physiological and evolutionary ecology. *A posteriori* methods provide useful catalogues and can be helpful in posing questions for further theoretical analysis, but leave the selective mechanisms behind the classification rather blurred. The shortcomings of this approach

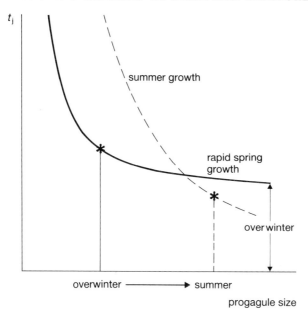

Fig. 7.6. Overwintering reproductive propagules should be small provided that spring growth is rapid, but larger propagules should be produced in early summer if summer growth is relatively slow. See text for details.

are, however, discussed in detail in Chapter 1. *A priori* systems of classification provide a framework for not only cataloguing data but also for rigorously testing predictions from the theory and as such attempt to bring the mechanisms more into focus. In this Chapter we have moved from the famous one-dimensional, *a priori*, r/K scheme of MacArthur to a two-dimensional, *a priori*, scheme based on aspects of the models discussed in Chapter 4. This classification incorporates the r/K dichotomy but also provides some insight into traits associated with various trophic niches, including ectoparasitism and endoparasitism and with aquatic oligotrophic and eutrophic systems.

As a final cautionary note, remember that associations between mortality, growth and birth rates can occur as a result of ecological compensation (p. 22). Hence in habitat comparisons observed differences might be due to this rather than to selection. The separation of genetic from purely ecological effects will require breeding studies under controlled conditions.

8

A More General Model

8.1 Introduction

A major goal of this book has been to define models that capture the essentials of (a) resource allocation and (b) its implications for fitness components S, n, and t. By combining these two aspects of organisms it is possible to make predictive statements about which allocation patterns (and hence dependent physiological, morphological, behavioural and life-cycle traits) should evolve in particular circumstances. The assumptions built into these models need to be critically evaluated, and this again involves making physiological observations. The predictions need rigorous testing, and refutations here will also normally lead, in the first instance, to a re-examination of assumptions and auxiliary hypotheses rather than to an attack on the core neo-Darwinian hypothesis. As is normal in science, this core hypothesis is a protected domain (Lakatos, 1978). Thus in all the foregoing chapters we have been attempting to sharpen our understanding of how organisms function, on the assumption that the fundamental premises of neo-Darwinism can be accepted. In this final chapter we attempt to push this programme further by making explicit a general class of models which most completely and generally captures the complexity of the functioning organism. These are based upon the PMP models that were introduced in Chapter 1 (Box 1.4) and developed further in Chapter 5. (PMP stands for Pontryagin's Maximum Principle).

8.2 The model

Assume food is to be allocated between several subsystems as in Fig. 8.1; these might be processes or structures. PMP (Box 1.4) shows that instant by instant the allocation of resources must maximize the weighted sum of birth rate and the growth rates of the various subsystems, minus mortality rate (Sibly & Calow, 1986). Thus finite resources should be channelled where they produce the greatest weighted growth

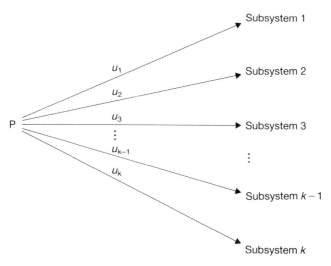

Fig. 8.1.   Allocation of the energy available for production (P) between k subsystems. A fraction u_1 is allocated to subsystem 1, u_2 to subsystem 2, and so on.

rates and birth rate, and minimize mortality rate. The weightings are all positive, and the allocation to a particular subsystem is weighted more if it has more effect on survivorship or birth rate.

Formally the rule is:

Maximize (birth rate) $\frac{1}{2}e^{-Ft}S_{t+}$ (growth rate of subsystem 1) λ_1 + (growth rate of subsystem 2) λ_2 ... + (growth rate of subsystem k) λ_k − (mortality rate) $\lambda_{k+1}S_t$

$$(8.1)$$

where $(\frac{1}{2}e^{-Ft}S_t)$, λ_1, λ_2, ... λ_k and λ_{k+1} represent the weightings given to the development of each subsystem and λ_1 ... λ_{k+1} are specified by differential equations (1.15) with boundary conditions (1.16). From now on we shall refer to expression 8.1 as the **Maximum Rule**.

8.3   Explanation of weighting factors

λ_{k+1} is the value of immediate self preservation in relation to immediate reproduction; i.e. it is a measure of the value of self (from the gene's point of view) in relation to the value of offspring. Thus resources should be allocated to self preservation if λ_{k+1} is high, but otherwise to reproduction. It can be appreciated intuitively that λ_{k+1} is closely related to Fisher's reproductive value (the number of young a female can expect to have over her whole life discounted by population growth rate). Formally λ_{k+1} = (reproductive value) $\times$ e^{-Ft} (Sibly *et al.*, 1985). λ_{k+1} is plotted as a function of age in Fig. 8.2 for a human being. Note that before the age of maturity, reproduction is impossible and therefore all resources should be devoted to growth and self preservation (according to the Maximum Rule given above). Assuming that the pay-off from allocating resources to reproduction does not vary with age, it follows from Fig. 8.2 that for humans self-preservation is most valuable in the years up to age 18, but thereafter declines steadily in value to reach zero at the age of the menopause (parental care has here been ignored). According to the present model, therefore, individuals should make increasing sacrifices to reproduce, reaching a maximum in their mid 40s.

In a similar way $\lambda_1, \ldots \lambda_k$ give the value of allocating resources to the body's various subsystems. Because they give the value of a subsystem in relation to the value (from a gene's point of view) of births they can be thought of as expressing the value of the subsystem in terms of birth-equivalents. In other words, if λ_{k+1} can be thought of as expressing the reproductive value of self-preservation then λ_i represents the reproductive value of allocating resources to subsystem i.

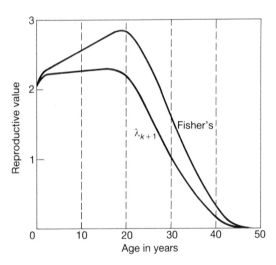

Fig. 8.2. The upper curve is Fisher's reproductive value, the lower curve is λ_{k+1}. The upper curve was calculated for Australian women in 1911, when $F = 0.01231$ per annum (after Fisher, 1958).

8.4 Application to various levels

The above principles can be applied at any level of organization. If the subsystems are at the cellular level this predicts that the allocation of resources should maximize the weighted sum of the growth rates of individual cells or cell populations (dependent on level of interest) and birth rate minus mortality rate of the organism containing the cells. In other words resources should be channelled where they are most effective in increasing the growth rates of the cellular systems and increasing the birth rate and reducing the mortality rate of the organism containing the cells.

Alternatively if the subsystems are 'total energy reserves', 'total protein reserves', and so on, the Maximum Rule is then concerned with the relative priorities of increasing different reserves. The maximized quantity is the weighted sum of birth rate and the rate of increasing energy reserves and the rate of increasing protein reserves and so on, minus mortality rate. Thus other things being equal the rate of increasing energy reserves (i.e. the net rate of obtaining energy) should be maximized. This is the basis of optimal foraging theory (Chapter 2). An example might be helpful. Suppose an animal had the choice of eating or drinking, i.e. adding to its energy or to its water reserves. Should it eat or drink? According to the Maximum Rule it should maximize the weighted sum of the rate of increase of energy and water reserves, minus mortality rate (assuming fecundity is zero). A worked example is given in Box. 8.1.

8.5 Application to differential organ growth

Assume that the subsystems are different organs with different functions (Fig. 8.3). Total body musculature is x_1 and the sizes of organs are x_2 to x_k. Food intake can be thought of as being dependent upon body musculature, because muscles are important in the capture of food and its subsequent manipulation and processing. Mortality rate (μ), on the other hand, will depend on both body musculature (e.g. active escape from predators) and the sizes of the other organs (because their functional ability will depend on their size, and the metabolic well-being of an organism depends upon the functioning of its individual organs — we shall be more explicit about this below). The growth rate of each organ ($\dot{x}_i$) will be proportional to the fraction (u) of resources allocated to it, so

$$\dot{x}_i = u_i P \text{ for } i = 1, \ldots k, \tag{8.2}$$

BOX 8.1. The choice between feeding and drinking (based on Sibly & McFarland, 1976).

Let x_1 = energy reserves, x_2 = water reserves, u_1 = fraction of time spent feeding, u_2 = fraction of time spent drinking. Thus $\dot{x}_1 = c_1 u_1$ and $\dot{x}_2 = c_2 u_2$, for some constants c_1 and c_2. The Maximum Rule states that feeding and drinking should maximize:—

$$\lambda_1 \dot{x}_1 + \lambda_2 \dot{x}_2 - \lambda_3 S_t \mu(x_1, x_2, u, u_2)$$

λ_1 and λ_2 specify the priorities that should be given to increasing energy reserves and water reserves respectively. Since $k = 2$, $\lambda_{k+1} = \lambda_3$ and this indicates the value of self preservation as explained in the text. μ is mortality rate and depends on four variables x_1, x_2, u_1 and u_2. Suppose these are related by the mathematical equation

$$\mu = (x_1 - \hat{x}_1)^2 + (x_2 - \hat{x}_2)^2 + u_1^2 + u_2^2$$

where $\hat{x}_1$ and $\hat{x}_2$ are constants specifying optimal values of energy and water reserves. The optimal solution is to eat and drink so that reserves increase with time according to a negative exponential curve, but if this is not possible because the animal is limited in the rate at which it can eat and drink, then it should feed if $\dot{x}_1(x_1 - \hat{x}_1) > \dot{x}_2(x_2 - \hat{x}_2)$, i.e. if (rate of feeding) × (food deficit) > (rate of drinking) × (water deficit), but it should drink in any other conditions. In fact these rules provided a fairly good description of the behaviour of Barbary doves in a series of operant experiments.

and expression (8.1) can be rewritten

$$H = \tfrac{1}{2} n e^{-Ft} S_t + \sum_{i=1}^{k} u_i P \lambda_i - \mu \lambda_{k+1} S_t. \qquad (8.3)$$

From equations (1.15), (1.16) and (8.3),

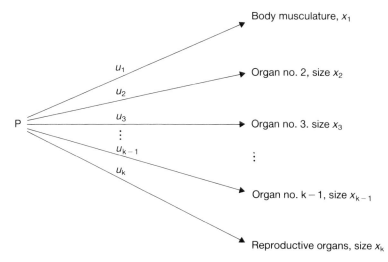

Fig. 8.3. Allocation of the energy available for production (P) between k organs. A fraction u_1 is alocated to body musculature, u_2 to organ 2, u_3 to organ 3, and so on.

$$\lambda_i(t) = -\int_t^\infty \frac{\partial \mu}{\partial x_i} \lambda_{k+1} S_t \, d\tau \text{ for i} = 2, \ldots k. \qquad (8.4)$$

It can be shown that $\lambda_2 \ldots \lambda_{k+1}$ are positive and eventually decline to zero (Sibly & Calow, 1986). It follows from equation (8.4) that the size of λ_i relative to $\lambda_2 \ldots \lambda_k$ only depends on $\frac{\partial \mu}{\partial x_i}$ (now and in the future) for i $= 2 \ldots k$, because the other term in equation (8.4), $\lambda_{k+1} S_t$, is the same for all the λs. So an organ for which $\frac{\partial \mu}{\partial x_i}$ is bigger (i.e. its size has more effect on survivorship) will have bigger λ_i — and so will receive greater priority under the optimal strategy. *Resources are therefore allocated between organs on the basis of their ability to reduce the mortality rate or improve the survivorship of the organism that contains them.*

This can be thought of as an **optimality rule for differential growth.** It makes explicit something that seems intuitively reasonable and were we able to measure the precise effects that different organs have on survivorship we should be able to make predictions about differential organ growth and hence about form. In the extreme case, for example, we can predict from equation (8.4) that if an organ has no effect on survivorship ($\frac{\partial \mu}{\partial x_i} = 0$) then it should receive no allocation, and this gives a basis for understanding the evolutionary loss of characters. For example, cave-dwelling fauna often lack eyes and pigment. There is no obvious survivorship cost to carrying these traits though they are useless in a completely dark environment, but since there is no gain then resources should not be wasted on them (Barr, 1968; but cf. Regal 1977 for an alternative explanation).

Not surprisingly, apart from these extreme cases, we do not have any precise measures of the $\frac{\partial \mu}{\partial x}$s, but we do have considerable information on differential organ growth. In principle we can work backwards from these data to the following questions:—

1 Are the observations compatible with the optimality rule for differential growth?

2 Do the observations allow us to make the optimality rule even more precise?

This procedure, of working backwards from observations to a specific optimality model, is known as the **inverse optimality programme** (Sibly & McFarland, 1976) and we shall pursue this further in the next section. Note that it tends towards the *a posteriori* approach discussed in Chapter 1, and in applying it we need to remind ourselves of an important theorem of systems analysis (Calow, 1976) that though a particular system (optimality model) always results in a particular and deducible behaviour, to a particular behaviour there can correspond a wide variety of different systems! So inverse optimality can never lead to a certain conclusion. Remember, though, that from the start we do have a particular class of models in mind (as hypotheses) and that we do not see inverse optimality as an end in itself, but rather as a preliminary way of sharpening our understanding of a complex system and of framing testable predictions about it.

8.6   Measuring risks; implications of allometry

Huxley (1932) discovered that the specific growth rate of an organ x_j usually stood in constant ratio to the specific growth rate of another organ or the whole organism (x_i):

$$\frac{\mathrm{d}x_j}{\mathrm{d}t} \times \frac{1}{x_j} : \frac{\mathrm{d}x_i}{\mathrm{d}t} \times \frac{1}{x_i} = \alpha:1. \tag{8.5}$$

Another way of interpreting this is that organ j receives α times more resources per unit weight than organ i; so α can be understood as a *distribution coefficient* indicating the share of resources received by organ j. Let us make this explicit and show if and how it relates to the optimality argument in the last section. Integrating equation (8.5) gives the famous allometric equation:

$$x_j = B \, x_i^{\alpha} \tag{8.6}$$

where B is a constant of integration.
Taking logarithms gives

$$\log x_j = \log B + \alpha \log x_i \tag{8.7}$$

Hence plotting $\log x_j$ against $\log x_i$ gives a straight line with a slope of α and this provides us with a way of determing α. Note that we are here concerned with intraspecific — even intra-individual — allometries, not the interspecific ones considered in Chapter 6. Nevertheless the usual statistical reservations apply to the determination of α (p. 110).

Fig. 8.4 shows some hypothetical, intraspecific relationships on both arithmetic (a) and logarithmic (b) coordinates. Huxley (1932) classified the patterns as follows:

$\alpha < 1$, j is in negative allometry to i
$\alpha = 1$, j and i are isometric
$\alpha > 1$, j is in positive allometry to i.

For $\alpha < 1$, relative to organ i, organ j grows fastest initially and later slowest; for $\alpha = 1$, the relative growth rate of the two organs is the same; for $\alpha > 1$, relative to organ i, organ j grows fastest late and slowest initially.

Fig. 8.5 shows organ weights of male rhesus monkeys plotted against body weight. For the gonads $\alpha > 1$, whereas for the heart $\alpha \sim 1$ and for the liver $\alpha < 1$, though the liver curve is not quite the same shape as the $\alpha < 1$ curve in Fig. 8.4a. According to the above argument the abscissa should be the weight of an organ, not the whole body. However since most of the body consists of body musculature and skeleton, the abscissa can loosely be thought of as 'body musculature', and in practice this would not introduce serious errors. The data in Fig. 8.5 were collected by Larson (1985) as part of a larger study of organ growth in five macaque and one baboon species. The distribution coefficients, α, (again relative to body weight) are displayed in Fig. 8.6. Overall it appears that $\alpha < 1$ for the heart, lungs, kidneys,

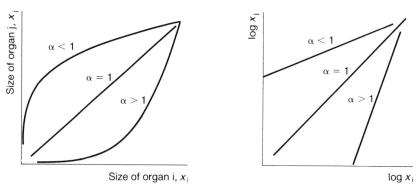

Fig. 8.4.   Growth of organ j relative to organ i. See text for discussion.

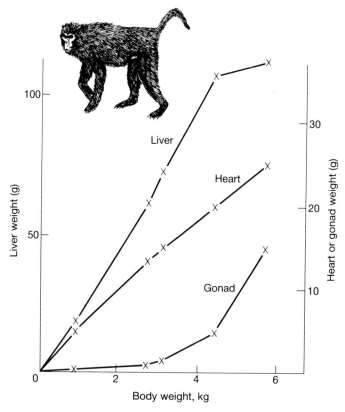

Fig. 8.5.   Organ weights plotted against body weight for male rhesus monkeys (*Macaca mulatta*) in the first five years of life. Redrawn from Larson (1985). See text for discussion.

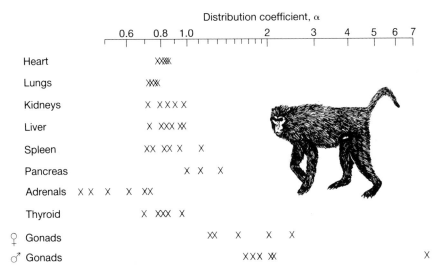

Distribution coefficient, α

	0.6	0.8	1.0	2	3	4	5	6	7

Heart — XXXX

Lungs — XXX

Kidneys — X XXXX

Liver — X XXX XX

Spleen — XX XX X   X

Pancreas — X X   X

Adrenals — X X   X   X   XX

Thyroid — X XXX X

♀ Gonads — XX   X   X   X

♂ Gonads — XXX XX   X

Fig. 8.6. Distribution coefficients for organ weights in relation to total body weight for six species of primate (redrawn from Larson 1985; some data points are missing). A log scale has been used for convenience. See text for discussion.

liver, spleen, adrenals and thyroid, whereas $\alpha > 1$ for the pancreas and gonads. This suggests that relative to total body weight the organs for which $\alpha < 1$ grow fastest initially and later slowest. Probably resources are allocated to these organs initially but are later diverted to body musculature as the animal approaches maturity.

What can we infer about mortality curves from a knowledge of the distribution coefficient α? It can be seen intuitively that the organs which obtain a large share of resources are likely to be those which contribute most to future survival. Remember that mortality depends on organ size — presumably if an organ is too small it cannot adequately fulfil its function, and the problem would get increasingly worse if the organ was progressively reduced in size. Hence a reasonable initial assumption is that the curve relating mortality μ to organ size x is convex seen from below, as in Fig 8.7. Different organs will have mortality curves of differing curvature. Intuition suggests that early in development resources should go to organs with initially steep, very convex mortality curves (e.g. Fig. 8.7a) and only subsequently to others (e.g. Fig. 8.7b). It should be noted that μ might start to increase with organ size after a critical size — if the organ gets too big for the body containing it, as in cancers. We are, however, ignoring this complication for the time being since: (a) we are working within the

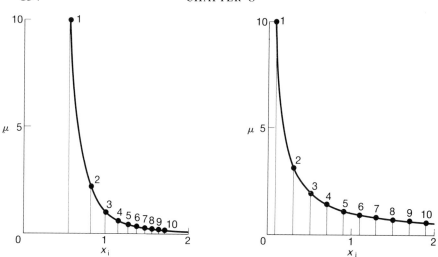

Fig. 8.7. Possible relationships between mortality rate and organ size (see text), numbers refer to age in days since birth. The optimal strategy is for organ i to grow faster than organ j early in life but for organ j to grow faster later. α for the relationship between j and i is 4, i.e. j is in positive allometry to i.

limits of normal growth and (b) it is difficult to know precisely how mortality would increase with size after the critical point — it seems very unlikely, for example, that $\mu(x)$ would be a symmetrical 'U'-shaped curve.

It turns out that the only conditions that are compatible with simultaneous organ growth are if the λs are equal at each age (Sibly & Calow, 1986). This means that the $\frac{\partial \mu}{\partial x}$s must be equal at each age, and so it follows that organs grow in such a way that the marginal increase of each makes the same contribution to reducing mortality. This can be considered as the *allometric version of the optimality rule for differential growth*. It is illustrated in Fig. 8.7 where for each age the slope for $\mu(x_i)$ equals the slope for $\mu(x_j)$. Suppose that these two mortalities are modelled by a rectangular hyperbola, and are independent of each other, i.e.

$$\mu = \left[\frac{c}{x_i^a} \right] + \left[\frac{d}{x_j^b} \right] \qquad (8.8)$$

Here the terms in the square brackets are formulae for inverse power functions and the plus sign represents the assumption of independence.

The slope of $\mu(x_i)$ is $-acx_i^{-a-1}$ and the slope of $\mu(x_j)$ is $-bdx_j^{-b-1}$ and as shown above these must be equal at each age. Therefore

$$-acx_i^{-a-1} = -bdx_j^{-b-1}$$

$$\therefore x_j = \left(\frac{bd}{ac}\right)^{\frac{1}{b+1}} x_i^{\frac{a+1}{b+1}} \tag{8.9}$$

Hence the allometric coefficient, α, equals $\dfrac{a+1}{b+1}$.

It follows that if $\alpha > 1$ the mortality curve for organ i is more convex and initially steeper than that for j, which is why it gets a greater share of resources early in life, as in Fig. 8.7.

This, then, is the explicit link, between the distribution of resources and optimality models, that we referred to at the start of this section. Where does it lead us? First, it does indicate that allometric growth is compatible with the optimality rule for differential growth. It should be noted that allometric growth is not compatible with all optimality models; for example some models predict discontinuities rather than continuous and simultaneous growth of different organs (Sibly & Calow, 1986). Second, the rule that organs grow in such a way that each makes the same contribution to reducing mortality, is a biologically reasonable basis for allometry. Third, in principle from the αs we should be able to make predictive statements about mortality curves and vice versa. Currently, we are only able to make what amount to rather extreme predictions, as for the loss of useless characters from cave-dwelling animals (p. 150) or rather vague ones such as, from the data in Fig. 8.6 it is to be expected that heart, lungs, kidneys and liver have more of an impact on mortality early in life than body musculature. All the former organs are concerned with metabolism whereas musculature is concerned with activity and hence behaviour. It is not unreasonable in mammals, with parental care, for metabolic organs to be more important for well-being (survival chances) early in life and for musculature and behaviour to become more important later in life; for example after release from parental care.

It should eventually be possible to test predictions more precisely than this by, for example, changing the mortality curves for particular organs whilst holding those for other organs constant (Sibly & Calow, 1986). It is possible to imagine both 'natural' and 'artificial' experiments here. For example, there is sometimes quite extensive variation within species between populations in defence structures — the thickness and sculpturing of snail shells and the spines on the backs of sticklebacks to name but two. Part of this might, according to the above argument, be related to the level of threat experienced by a particular population; in particular there should be a predictable and precise correlation between the size of the defence structure and the intensity of predation. Similarly, it ought to be possible to alter the extent to which laboratory populations are subjected to stressors of various kinds. Thus the liver is a major site of detoxification of various pollutants; how and to what extent is its size modulated by the level of various toxins?

There are, of course, complications that we have ignored. In particular we have said nothing about the extent to which developmental constraints rather than the mortality relationships are likely to influence the αs. For example, to what extent does morphogenesis put limits on when and to what extent organs can develop in order to fill available space within the developing embryo? These are likely to be important but, *a priori*, it is difficult to be precise about how important. As is usual, the only way of sorting out the relative contribution of constraints and optimization is on the basis of observations — ideally, designed to test predictions from conflicting models. This is likely to be an interesting and lively area of future research.

9
General Summary

This book applies the optimality approach to animal physiology. Because evolution is an optimizing process, it follows that if we can calculate the rates of increase of genes which affect life histories in specified ways, then we can identify those genes which spread fastest, and are therefore likely to be final products of the evolutionary process. We refer to the rate of spread as fitness, and paraphrase the above by saying that evolution maximizes fitness. The general methods of calculating fitness are introduced in Chapter 1 in terms of the key life-history variables, namely the rates of reproduction, mortality and growth, at different ages. In general genes that maximize rates of reproduction and growth and minimize mortality rate will be favoured — unless there are trade-offs between key life-history variables so that improvement of one necessarily involves deterioration of another. The exact form of these trade-offs depends on the details of the animal's physiology, and we see their precise delineation as being one of the central challenges for physiological ecology.

One major cause of trade-offs lies in the laws of conservation of energy and matter, from which it follows that energy/matter allocated to one function (e.g. reproduction) is denied to others (e.g. growth). We call this the Principle of Allocation, and it forms the basis of allocation diagrams throughout the book. The simplest case, of allocation between growth and reproduction, is illustrated in Fig. 1.2, and the argument is extended to include defence (against mortality agents) in Fig. 1.4, although the full analysis of this case has to wait until Chapter 5. Chapter 3 considers the allocation of energy between maintenance and production (Fig. 3.1), and the analysis reaches its most complete form in considering the allocation of resources between all the organs of the body in Chapter 8.

However physiological constraints and trade-offs can also arise from causes other than the Principle of Allocation. For example there can be negative correlations between physiological rates and efficiencies, and these are examined in Chapters 2 and 3. Links between physiological rates and survival chances are discussed in Chapters 4, 5, 7 and 8.

Chapter 2 begins with a consideration of optimal foraging theory, which is historically apposite because many of the ideas and techniques of biological optimization theory were first developed in the context of foraging in the 1970s. The link with fitness arises because animals that maximize their input of energy can breed sooner and can invest more in reproduction and defence. Although, for the sake of argument, posed in terms of energy it is understood that this will not always be the right currency; so, for example, the argument would have to be reformulated if sodium were the limiting factor (Fig. 2.2). Specific predictions are made about which items should be included in the optimal diet and which should be ignored, and if food is distributed in patches which become depleted by foraging, then predictions are made about how long to remain in one patch before travelling on to the next. Mathematical models of this type can also be applied to a detailed study of digestion to make predictions about how long digesta should be retained in the absorptive regions of the gut, and to explain why continuous feeders often have larger guts when feeding on poorer foods (Table 2.1). The theory can also explain why deposit feeders selectively feed from small rather than large sediment particles (Fig. 2.6). Characteristic of these models is a decline in digestive efficiency with increasing food intake.

Chapter 3 uses the same approach to give an overview of the way energy is used in the body. Negative relationships between the rates and efficiencies of physiological processes may be very general (e.g. Fig. 3.3), so that although there is a general advantage to efficient processes, there may be a place also for less efficient processes which occasionally operate at a higher rate, allowing the animal a sudden burst of speed, for example, to escape predators or catch prey.

Because of the Principle of Allocation, energy and matter are divided between production (growth and reproduction) and maintenance (i.e. continued survival) (Fig. 3.1). It follows from the negative relationship between digestive efficiency and food intake (examined in Chapter 2) that there should be a curvilinear relationship between production and food intake, and this idea is supported by data from agricultural research (Fig. 3.4c). This sort of understanding of the costs of production can also give insight into the evolution of different foraging patterns and of homeothermy. Costs of maintenance cause reduced growth rates and/or reproductive rates, but can bring substantial survival gains.

The evidence for trade-offs is very briefly reviewed at the end of Chapter 3, and some important studies not discussed later are illustrated here. Methods of demonstrating costs and identifying trade-offs are considered in detail, and some pitfalls are pointed out. Despite being based on fundamental physical and chemical constraints these trade-offs have proved surprisingly difficult to expose. Experimentally, one usually hopes to map them by changing one variable and observing the response of others and, in principle, this may be achieved by physiological (e.g. hormone) treatment, clutch size manipulation, breeding experiments and genetical selection.

Chapter 4 focuses on trade-offs involving investment in reproduction. Maximum insight is gained from the simplest model that captures the essentials of the processes involved, and in Chapter 4 it is therefore assumed that growth stops at the age of first reproduction, and that the same reproductive strategy is used at each breeding attempt. Thus the key life-history variables of mortality rate and the timing of reproduction may differ between juveniles and adults but not within each category. Hence the life cycle can be characterized by five variables describing time to first breeding (t_j), mortality rate during this period (μ_j), fecundity (n), time between breeding (t_a) and adult mortality rate (μ_a). It is very likely that these key variables can be traded-off against each other, the classic example being the 'cost of reproduction' which involves a survival cost to the adult if fecundity is increased. If the form of this latter trade-off is known (as in Fig. 4.4) then predictions can be made about how much to invest in reproduction. It turns out that animals should invest more the greater the chance the juveniles will survive. Thus in comparing populations which differ only in the mortality experienced by juveniles, it is to be expected that the population experiencing the greater mortality should invest less in reproduction. This illustrates what can be achieved by detailed analysis of a single trade-off. In fact with five key life-history variables there are ten possible trade-offs (taking the variables a pair at a time, as in Table 4.1), and seven of these are considered in Chapter 4. In addition to the cost of reproduction two trade-offs are especially emphasized. One concerns the trade-off between number and survivorship of offspring, which was first suggested and experimentally analysed by David Lack in the 1940s, and has been historically important in the development of Evolutionary Biology. The prediction here is that clutch size will maximize the number of young surviving to breed, and this is very successful in explaining clutch sizes (Table 4.2), although some clutches are slightly less than predicted. The second trade-off to which attention is drawn involves n and t_j and arises because more offspring can be produced if each is allocated fewer resources. But if gametes are less well provisioned they take longer to grow up, other things being equal. The prediction here is that smaller eggs should be produced in environments where juvenile growth rates are higher.

Some complications arise in this chapter. One which we call the Principle of Compensation states that it is not possible for stable populations to differ in only one life-history variable. For example if juvenile mortality rate is worse in population A than in population B then the numbers of population A must decline unless some other life-history variable compensates, so that for example adult mortality rate or fecundity is better. Other complications arise because a different model is appropriate if the costs of reproduction are paid before instead of after reproduction (Fig. 4.1), and because the environment may vary from year to year - - an important area of investigation only briefly touched on here.

In Chapter 5 attention is shifted from reproduction to growth. The basis of the argument is that juvenile growth rates may be held at submaximal levels because growing faster increases mortality costs. A mathematical treatment shows that sigmoid growth curves, which are very widely observed in nature (examples in Fig. 5.2), can only be optimal strategies if mortality rate decreases with body size, and the mortality costs of growing faster increase increasingly with growth rate (Table 5.1). If the trade-offs could be accurately measured the predictions would be precise and quantitative, but to date the supporting evidence for them is only circumstantial. Optimality models can also make predictions about when animals should stop growing and start breeding. The circumstances favouring a sharp switch are broader than those favouring a progressive one. Again there is an acute need for data on the distribution of these life-cycle types and on the demographies and trade-offs with which they are associated. A brief exposition is included of an alternative class of models in which growth rates are always maximal, at physiological/developmental limits, and sigmoid growth curves are generated because of changes in constraints with age.

Chapter 6 considers the manifold effects and implications of body size. Body size affects the speed and efficiency of many physiological processes and has many implications for rates of growth, mortality and reproduction. For example at a very simplistic level it takes longer to grow to be bigger, big animals tend to be less liable to mortality than small, and large animals are more fecund than smaller conspecifics. However because there are so many ramifications of body size it is not easy to identify the resulting effect on fitness. The approach taken here is to begin by considering a number of constraints on physiological processes imposed by geometric and physical factors, and to ask whether the contraints are achieved or not, using interspecific allometric data. It appears that minimal and maximal metabolic rate, for inactive and maximally active animals respectively, may be governed by size-dependent constraints, and consideration of maximum activity levels gives some understanding of leg muscle morphology in large animals. Several key life-history parameters, however, do not lie on the minimum or maximum constraints imposed by body size. The efficiency of some physiological processes is shown to decline with body size, at least when animals with the same body plan are compared, and this may be a result of morphological constraints. If the overall effect of body size on fecundity and on age at maturity is known then precise predictions can be made about optimal body size (Fig. 6.10).

In Chapter 7 we return to some of the predictions derived in Chapter 4, and use them to produce a two-way classification of habitats (Table 7.3) in terms of habitat effects on juvenile growth rate, G, and on a survivorship index, S. This classification incorporates the r/K dichotomy as a special case, and seems to be widely successful in predicting differences in egg size associated with differences in temperature and food supply which together determine growth rate. What circumstantial evidence there is relating to differences in survivorship is also generally supportive.

Chapter 8 represents a first attempt at a general theory of organ function in relation to organ size and growth. Most of the principles and methods developed in the book here reach their most sophisticated form. According to the Principle of Allocation the resource input is divided between all the organs of the body, and investment in each should be related to its ability to reduce mortality rate. The allometric curves for organ growth during development are shown to be related to the organ size—mortality curves. Organs with fast early development should have more convex and initially steeper mortality curves than those developing later (Fig. 8.7).

The advantage of the optimality approach advocated in this book is that it provides a unified treatment of a wide range of physiological problems, and sets each in its ecological and evolutionary context. We are well aware that it is hard to evaluate the theory fully because of a lack of data in some areas. Our hope is that this book will encourage experimental work to plug some of the gaps.

Glossary

Page numbers where each symbol is introduced are given in brackets.

A **variable** is a quantity of interest, variation in which is explicitly considered. A **parameter** is a variable which is being held constant at present. It is sometimes useful to refer to a collection of variables as a **vector** variable (e.g. **x**), and single variables are then called **scalar** variables.

A **mathematical function** is a curve, of the sort illustrated in many of the figures in this book. More formally it is a particular type of relationship between two variables, e.g. x and y, such that y is uniquely defined for every value of x: the notation $y(x)$ is used to signify that y is a mathematical function of x. A **monotonic increasing** function $y(x)$ is a function such that $x_1 > x_2 \Rightarrow y(x_1) > y(x_2)$. $\Rightarrow$ means 'implies'. A **monotonic decreasing** function $y(x)$ is a function such that $x_1 > x_2 \Rightarrow y(x_2) < y(x_1)$. The functions used in this book are tacitly assumed to be continuous and differentiable.

*	used to denote an optimal strategy (i.e. a strategy maximizing fitness) (p. 15)
α	allometric exponent (p. 150)
a, b, c, d	constants
A	absorbed energy per unit time (watts) (p. 45)
BMR	basal metabolic rate
C_a	energy costs of activity per unit time (watts) (p. 53)
C_m	energy costs of maintenance per unit time (watts) (p. 45)
C_p	energy costs of production per unit time (watts) (p. 45)
e	base of natural logarithms, 2.718 ...
$f(x)$	a function of x
F	fitness, defined as per capita rate of increase of a gene in a specified environment (p. 10)
F	force (Chapter 6)
$g(x)$	a function of x
G	a habitat index of growth rate (p. 133)

H	the Hamiltonian. Moment by moment maximization of H leads to maximum fitness, so in a sense H is a fitness measure. (p. 20)
I	food input (watts) (p. 45)
k	efficiency of a process, i.e. output/input. The main efficiencies are summarized in Fig. 3.1 (p. 45)
k_1	gross production efficiency, i.e. P/I (p. 45)
k_b	biochemical efficiency (p. 46)
k_d	digestive efficiency, i.e. the net energy obtained from 1 g of food after digesting it for time t (p. 35)
K	a constant. Used for carrying capacity of the population in Chapter 7
λ	a variable used in Pontryagin's Maximum Principle (p. 21)
m	body mass (p. 19)
μ_a	per capita mortality rate of adults (p. 70)
μ_j	per capita mortality rate of juveniles (p. 70)
n	number of offspring per breeding attempt. If this changes with time, n_t denotes the number of offspring produced per unit time by an adult of age t. (p. 11, 66)
N	number of individuals in the population (p. 127)
P	rate of production of new tissue (watts) (p. 45)
P_r	amount of resource available for producing and provisioning gametes. $P_r = n \times z$. (p. 84)
r	radius of leg (Chapter 6)
r	rate of population increase when population density is very low (Chapter 7)
R	a constant
S	used generally for survivorship variables except in Section 5.6 and Chapter 7
S	an index of survivorship (Chapter 7) (p. 131)
S_a	adult survivorship between breeding attempts (p. 66)
S_j	juvenile survivorship from birth until first breeding (p. 66)
S_t	survivorship from birth until age t (p. 11)
t, T, τ	various time variables. t is used for age in Chapter 1.
t_a	interval between breeding attempts (p. 66)
t_d	retention time. The time from starting digestion that digesta are retained in the absorptive regions of the gut (p. 33)
t_j	age at first breeding (p. 66)
u	a control variable (p. 20)
Φ	a phenotypic measure of fitness (p. 20)
v	rate at which food items enter the digestive chambers (p. 33)

x, y	variables. x is used for state variables in Chapters 1 and 8
z	offspring size (p. 84)

The usage of f, H, λ, $\mathbf{u}$ and $\mathbf{x}$ follows that of optimal control theory. k_1 and k_2 follows Ivlev (1961), Blaxter (1980), and most treatments of animal production. r and K in Chapter 7 follow the conventions of Population Ecology.

References

Abrahamson, W.G. & Caswell, H. (1982) On the comparative allocation of biomass, energy and nutrients in plants. *Ecology*, **63**, 982–991.

Alexander, R. McN. (1982) *Optima for Animals*. Edward Arnold, London.

Alexander, R. McN., Jayes, A.S., Maloiy, G.M.O. & Wathuta, E.M. (1979) Allometry of the limb bones of mammals from shrews (*Sorex*) to elephant (*Loxodonta*). *Journal of Zoology*, London, **189**, 305–314.

Alexander, R. McN., Jayes, A.S., Maloiy, G.M.O. & Wathuta, E.M. (1981) Allometry of the leg muscles of mammals. *Journal of Zoology*, London, **194**, 439–552.

Anderson, R. A. & Karasov, W.H. (1981) Contrasts in energy intake and expenditure in sit-and-wait and widely foraging lizards. *Oecologia* (Berlin), **49**, 67–72.

Askenmo, C. (1979) Reproductive effort and return rate of male pied flycatchers. *American Naturalist*, **114**, 748–752.

Barr, T.C. (1968) Cave ecology and the evolution of troglobites. *Evolutionary Biology*. (Eds T. Dobzhansky, M.K. Hecht & W.C. Steere), vol. 2, pp. 35–102. Appleton-Century, Crofts, New York.

Bayne, B.L. (Ed.) (1976) *Marine Mussels: Their Ecology and Physiology*. Cambridge University Press, Cambridge, UK.

Bayne, B.L. (1986) Genetic aspects of adaptation in bivalve molluscs. *Evolutionary Physiological Ecology*. (Ed. P. Calow). Cambridge University Press, Cambridge, UK.

Bell, G. (1982) *The Masterpiece of Nature*. Croom Helm, London & Canberra.

Bell, G. (1984a) Measuring the cost of reproduction. 1. The correlation structure of the life table of a planktonic rotifer. *Evolution*, **38**, 300–313.

Bell, G. (1984b) Measuring the cost of reproduction. 2. The correlation structure of the life tables of five freshwater invertebrates. *Evolution*, **38**, 314–326.

Belovsky, G.E. (1978) Diet optimization in a generalist herbivore: the moose. *Theoretical Population Biology*, **14**, 105–134.

Bertalanffy, L. von (1960) Principles and theory of growth. *Fundamental Aspects of Normal and Malignant Growth* (Ed. W. Nowinski) pp 137–252. Elsevier, Amsterdam.

Beverton, R.J.H. & Holt, S.J. (1959) A review of the lifespans and mortality rates of fish in nature and their relation to growth and other physiological characteristics. In CIBA Foundation Colloquia on Ageing, Vol. 5. *The Lifespan of Animals*. (Eds G.E.M. Wolstenholme & C.M. O'Connor) pp 142–180. J. & A. Churchill, London.

Blaxter, K. (1980) *The Nutrient Requirements of Ruminant Livestock*. Technical Review by an Agricultural Research Council Working Party. Commonwealth Agricultural Bureaux, Slough, England.

Boucher-Rodoni, R. (1973) Vitesse de digestion d'*Octopus cyanaea* (Cephalopoda: Octopoda). *Marine Biology*, **18**, 237–242.

Boyce, M.S. (1984) Restitution of r- and K-selection as a model of density-dependent natural selection. *Annual Review of Systematics and Ecology*, **15**, 427–441.

Brattsten, L.B., Wilkinson, C.F. & Eisner, T. (1977) Herbivore-plant interactions: mixed function oxidases and secondary plant substances. *Science*, **196**, 1349–1352.

Brody, S. (1945) *Bioenergetics and Growth*. Reinhold Publ. Corp. N.Y.

Bryant, D.M. (1979) Reproductive costs in the house martin. *Journal of Animal Ecology*, **48**, 655–675.

Buckland, S.T. (1982) A mark-recapture survival analysis. *Journal of Animal Ecology*, **51**, 833–847.

Calder, W.A. (1984) *Size, Function and Life History*. Harvard University Press, Cambridge Mass.

Calow, P. (1973) The relationship between fecundity, phenology and longevity: a systems approach. *American Naturalist*, **107**, 559–574.

Calow, P. (1976) *Biological Machines. A Cybernetic Approach to Life*. Edward Arnold, London.

Calow, P. (1977a) Ecology, evolution and energetics: a study in metabolic adaptation. *Advances in Ecological Research*, **10**, (Ed. A. Macfadyen), pp 1–61. Academic Press, London, New York.

Calow, P. (1977b) Conversion efficiencies in heterotrophic organisms. *Biological Reviews*, **52**, 385–409.

Calow, P. (1979) The cost of reproduction — a physiological approach. *Biological Reviews*, **54**, 23–40.

Calow, P. (1981) *Invertebrate Biology. A Functional Approach*. Croom Helm, London.

Calow, P. (1983a) *Evolutionary Principles*. Blackie, Glasgow and London.

Calow, P. (1983b) Life-cycle patterns and evolution. *The Mollusca*, **6**, (Ed. W.D. Russell-Hunter) 649–678. Academic Press, New York.

Calow, P. (1983c) Pattern and paradox in parasite reproduction. *Parasitology*, **86**, 197–207.

Calow, P. (1983d) Energetics of reproduction and its evolutionary implications. *Biological Journal of the Linnean Society*, **20**, 153–165.

Calow, P. (1984) Exploring the adaptive landscapes of invertebrate life cycles. *Advances in Invertebrate Reproduction*, **3**, 329–342.

Calow, P. (1985) Adaptive aspects of energy allocation. *Fish Energetics, New Perspectives* (Eds. P. Tytler & P. Calow) pp 13–31. Croom Helm, London & Sydney.

Calow, P. (in press) Fact and Theory — an overview. *Cephalopod Life Cycles, Vol 2* (Ed. P. Boyle) Academic Press, London.

Calow, P., Beveridge, M. & Sibly, R. (1979) Adaptational aspects of asexual reproduction in freshwater triclads. *American Zoologist*, **19**, 715–727.

Calow, P. & Riley, H. (1982) Observations on reproductive effort in British erpobdellid and glossiphoniid leeches with different life cycles. *Journal of Animal Ecology*, **51**, 697–712.

Calow, P. & Sibly, R. (1983) Physiological trade-offs and the evolution of life cycles. *Science Progress*, Oxford, **68**, 177–188.

Calow, P. & Townsend, C.R. (1981a) Energetics, ecology and evolution. *Physiological Ecology. An Evolutionary Approach to Resource Use* (Eds C.R. Townsend & P. Calow), pp 3–19. Blackwell Scientific Publications, Oxford.

Calow, P. & Townsend, C.R. (1981b) Resource utilization in growth. *Physiological Ecology. An Evolutionary Approach to Resource Use*. (Eds C.R. Townsend & P. Calow), pp 220–244. Blackwell Scientific Publications, Oxford.

Calow, P. & Woollhead, A.S. (1977) The relationship between ration, reproductive effort and age-specific mortality in the evolution of life-history strategies — some observations on freshwater triclads. *Journal of Animal Ecology*, **46**, 761–781.

Caraco, T., Martindale, S. & Pulliam, H.R. (1980) Flocking: advantages and disadvantages. *Nature*, London, **285**, 400–401.

Case, T.J. (1978) On the evolution and adaptive significance of postnatal growth rates. *Quarterly Review of Biology*, **53**, 243–279.

Cathcart, E.P. (1953) The early development of the science of nutrition. *Biochemistry and Physiology of Nutrition* Vol 1 (Eds G.F. Bourne & G.W. Kidder), pp. 1–16. Academic Press, New York.

Causton, D.R. (1983) *A Biologist's Basic Mathematics*. Edward Arnold, Baltimore.

Charlesworth, B. (1980) *Evolution in Age-Structured Populations*. Cambridge University Press, Cambridge.

Charnov, E.L. (1982) Parent-offspring conflict over reproductive effort. *American Naturalist*, **119**, 736–737.

Chia, F.S. (1976) Sea anemone reproduction: patterns and adaptive radiations. *Coelenterate Behaviour and Ecology* (Ed. G.O. Mackie) pp 261–270. Plenum Press, New York.

Christiansen, F.B. & Fenchel, T.M. (1979) Evolution of marine invertebrate reproductive patterns. *Theoretical Population Biology*, **16**, 267–282.

Clark, A. (1979) On living in cold water: K strategies in Arctic benthos. *Marine Biology*, **55**, 111–119.

Clutton-Brock, T.H. (1984) Reproductive effort and terminal investment in iteroparous animals. *American Naturalist*, **123**, 212–229.

Clutton-Brock, T.H. & Harvey, P.H. (1979) Comparison and adaptation. *Proceedings of the Royal Society of London*, **205**, 547–565.

Clutton-Brock, T.H. & Harvey, P.H. (1983) The functional significance of variation in body size in mammals. *Mammal Behaviour and Ecology*. (Ed. J.F. Eisenberg) pp 632–663. Smithsonian Institute, Washington D.C.

Clutton-Brock, T.H., Guinness, F.E., & Albon, S.D. (1982) *Red Deer*. University of Chicago Press, Chicago.

Clutton-Brock, T.H., Guinness, F.E. & Albon, S.D. (1983) The costs of reproduction to red deer hinds. *Journal of Animal Ecology*, **52**, 367–383.

Cody, M.L. (1966) A general theory of clutch size. *Evolution*, **20**, 174–184.

Cody, M.L. (1973) Coexistence, coevolution and convergent evolution in seabird communities. *Ecology*, **54**, 31–44.

Cowie, R.J. (1977) Optimal foraging in great tits, *Parus major*. *Nature*, London, **268**, 137–139.

Davies, P.S. (1966) Physiological ecology of *Patella*. 1. The effect of body size and temperature on metabolic rate. *Journal of Marine Biological Association*, U.K., **46**, 647–658.

Davies, P.S. (1969) Physiological ecology of *Patella*. 111. Desiccation effects. *Journal of Marine Biological Association*, U.K., **49**, 291–304.

Dawkins, M. (1971) Perceptual changes in chicks: another look at the "search image" concept. *Animal Behaviour*, **19**, 46–54.

Dawkins, R. & Carlisle, T.R. (1976) Parental investment, mate desertion and a fallacy. *Nature*, **262**, 131–132.

Deevey, E.S. (1947) Life tables for natural populations of animals. *Quarterly Review of Biology*, **22**, 283–314.

Demment, M.W. & Soest, P.J. van (1985) A nutritional explanation for body-size patterns of ruminant and nonruminant herbivores. *American Naturalist*, **125**, 641–672.

Drent, R.H. & Daan, S. (1980) The prudent parent: energetic adjustments in avian breeding. *Ardea*, **68**, 225–252.

Ebert, T.A. (1985) Sensitivity of fitness to macroparameter changes: an analysis of survivorship and individual growth in sea urchin life histories. *Oecologia*, **65**, 461–467.

Eisenberg, J.F. (1981) *The Mammalian Radiations*. Athlone Press, London.

Enckell, P.H. & Nilsson, L.M. (1985) Eds. Plant-animal interactions. *Oikos*, **44**, 1–228.

Falconer, D.S. (1981) *Introduction to Quantitative Genetics* (second Edn.) Longman, London & New York.

Fenchel, T. (1974) Intrinsic rate of natural increase: the relationship with body size. *Oecologia*, (Berlin) **4**, 317–376.

Fisher, R.A. (1930) *The Genetical Theory of Natural Selection*. The Clarendon Press, Oxford. 2nd edition 1958, Dover, New York.

Fox, H.M. & Simmonds, B.G. (1933) Metabolic rates of aquatic arthropods from different habitats. *Journal of Experimental Biology*, **10**, 67–74.

Gadgil, M. & Bossert, W.H. (1970) Life historical consequences of natural selection. *American Naturalist*, **104**, 1–24.

Gnaiger, E. (1983) Heat dissipation and energetic efficiency in animal anoxibiosis: economy contra power. *Journal of Experimental Zoology*, **228**, 471–490.

Gnaiger, E. (1986) Optimum efficiencies of energy transformation in anoxic metabolism, the strategies of power and economy. *Evolutionary Physiological Ecology*. (Ed. P. Calow). Cambridge University Press, Cambridge, UK.

Goodman, D. (1984) Risk spreading as an adaptive strategy in iteroparous life histories. *Theoretical Population Biology*, **25**, 1–20.

Gould, S.J. (1966) Allometry and size in ontogeny and phylogeny. *Biological Reviews*, Cambridge, **41**, 587–640.

Greenslade, P.J.M. (1983) Adversity selection and the habitat templet. *American Naturalist*, **122**, 352–365.

Grime, J.P. (1974) Vegetation classification by reference to strategies. *Nature*, London, **250**, 26–31.

Grime, J.P. (1979) *Plant Strategies and Vegetation Processes*. Wiley, Chichester.

Gutschick, V.P. (1981) Evolved strategies in nitrogen acquisition by plants. *American Naturalist*, **118**, 607–637.

Haldane, J.B.S. (1927) *Possible Worlds*. Chatto, London.

Hamilton, W.D. (1966) The moulding of senescence by natural selection. *Journal of Theoretical Biology*, **12**, 12–45.

Henneman, W.W. (1983) Relationship among body mass, metabolic rate and intrinsic rate of natural increase in mammals. *Oecologia*, (Berlin), **56**, 104–110.

Hunt, R. (1982) *Plant Growth Curves*. Edward Arnold, London.

Hussell, D.T.J. (1972) Factors affecting clutch size in arctic passerines. *Ecological Monographs*, **49**, 317–364.

Huxley, J.S. (1932) *Problems of Relative Growth*. Methuen, London.

Itô, Y. (1980) *Comparative Ecology* (2nd Edn., Transl. by J. Kikkawa). Cambridge Univ. Press, Cambridge.

Ivlev, V.S. (1961) *Experimental Ecology of the Feeding of Fishes*. Yale University Press, New Haven.

Jackson, C.M. (1937) Recovery of rats upon refeeding after prolonged suppression of growth by underfeeding. *Anatomical Record*, **68**, 371–381.

Janzen, D.H. (1981) Evolutionary physiology of personal defence. *Physiological Ecology: an evolutionary approach to resource use* (Eds C.R. Townsend & P. Calow), pp 145–164. Blackwell Scientific Publications, Oxford.

Jones, M.B. (1978) Aspects of the biology of the big-handed crab, *Heterozius rohendifrons* (Decapoda: Brachyura), from Kaikoura, New Zealand. *New Zealand Journal of Zoology*, **5**, 783–794.

Jungers, W.L. (1985) Ed. *Size and Scaling in Primate Biology*. Plenum, New York.

King, J.L. (1984) Natural selection and blending inheritance. *Evolutionary Theory*, **7**, 41–43.

Kleiber, M. (1933) Tiergrobe und Futterverwertung. *Tierernahrung*, **5**, 1–12.

Klomp, H. (1970) The determination of clutch size in birds: a review. *Ardea*, **58**, 1–124.

Kolding, S. & Fenchel, T.M. (1981) Patterns of reproduction in different populations of five species of the amphipod genus *Gammarus*. *Oikos*, **37**, 167–172.

Krebs, J.R. & McCleery, R.H. (1984) Optimization in behavioural ecology. *Behavioural Ecology: An evolutionary approach* (2nd Edn.) (Eds. J.R. Krebs & N.B. Davies), pp 91–121. Blackwell Scientific Publications, Oxford.

Lack, D. (1954) *The Natural Regulation of Animal Numbers*. Clarendon Press, Oxford.

Lack, D. (1966) *Population Studies of Birds*. Clarendon Press, Oxford.

Lack, D. (1968) *Ecological Adaptation for Breeding in Birds*. Methuen, London.

Lakatos, I. (1978) *The Methodology of Scientific Research Programmes*. (Eds T. Worrall & G. Currie). Cambridge University Press, Cambridge, UK.

Lam, R.K. & Frost, B.W. (1976) Model of copepod filtering response to changes in size and concentration of food. *Limnology and Oceanography*, **21**, 490–500.

Lang, J. (1981) The nutrition of the commercial rabbit. *Nutrition Abstracts and Reviews — Series B*, **51**, 197–225.

Larson, S.G. (1985) Organ weight scaling in primates. *Size and Scaling in Primate Biology* (Ed. W.L. Jungers), pp 91–113. Plenum, New York.

Lawton, J.H. (1970) Feeding and food energy assimilation in larvae of the damselfly *Pyrrhosoma nymphula* (Sulz.) (Odonata: Zygoptera). *Journal of Animal Ecology*, **39**, 669–689.

Lehman, J.T. (1976) The filter feeder as an optimal forager and the predicted shapes of feeding curves. *Limnology and Oceanography*, **21**, 501–516.

Lehninger, A.L. (1973) *Bioenergetics* W.A. Benjamin, Inc., Menlo Park, Calif. (2nd Edn.)

Leutenegger, W. (1976) Allometry in neonatal size in eutherian mammals. *Nature London*, **263**, 229–230.

Lewin, R. (1982) How did humans evolve big brains? *Science*, **216**, 840–841.

Lewontin, R.C. (1965) Selection for colonizing ability. *The Genetics of Colonizing Species*. (Eds. H.G. Baher & G.L. Stebbins) pp 77–94. Academic Press, New York.

Lewontin, R.C. (1978) Adaptation. *Scientific American*, **239**, 156–165.

Lindeman, R.L. (1942) The trophic-dynamic aspects of ecology. *Ecology*, **23**, 399–418.

Livingstone, D.R. (1983) Invertebrate and vertebrate pathways of anaerobic metabolism: evolutionary considerations. *Journal of the Geological Society*, **140**, 27–38.

Luckinbill, L.S., Arking, R. & Clare, M.J. (1984) Selection for delayed senescence in *Drosophila melanogaster*. *Evolution*, **38**, 996–1003.

MacArthur, R.H. (1972) *Geographical Ecology*. Harper & Row, New York.

McCance, R.A. & Widowson, E.M. (1962) Nutrition and growth. *Proceedings of the Royal Society (B)*, **156**, 326–337.

McClure, P.A. & Randolph, J.C. (1980) Relative allocation of energy to growth and development of homeothermy in the eastern wood rat (*Neotoma floridiana*) and tropical cotton rat (*Sigmodon hispidus*). *Ecological Monographs*, **50**, 199–219.

McMahon, T. & Bonner, J.T. (1983) *On Size and Life*. Scientific American Books, New York.

McNab, B.K. (1980) Food habits, energetics and the population biology of mammals. *American Naturalist*, **116**, 106–114.

Maloiy, G.M.O., Alexander, R.McN., Njau, R. & Jayes, A.S. (1979) Allometry of the legs of running birds. *Journal of Zoology*, London, **187**, 161–167.

Maltby, L. & Calow, P. (1986) Intraspecific life-history variation in *Erpobdella octoculata* (Hirudinea, Erpobdellidae). II. Testing theory on the evolution of semelparity and iteroparity. *Journal of Animal Ecology*, **55** (in press).

Martin, R.D. (1981) Relative brain size and metabolic rate. *Nature*, London, **293**, 57.

Mayer, T. (1948) Gross efficiency of growth of the rat as a simple mathematical function of time. *Yale Journal of Biological Medicine*, **21**, 415–419.

Maynard Smith, J. (1978) *The Evolution of Sex*. Cambridge University Press, Cambridge.

Maynard Smith, J. (Ed.) (1985) *On Being the Right Size and Other Essays*. Oxford University Press, Oxford.

Meats, A. (1971) The relative importance to population increase of fluctuations in mortality, fecundity and the time variables of the reproductive schedule. *Oecologia*, **6**, 223–237.

Nelson, J.B. (1964) Factors influencing clutch-size and chick growth in the North Atlantic Gannet, *Sula bassana. Ibis*, **106**, 63–77.

Nordwijk, A.J. Van. (1984) Quantitative genetics in natural populations of birds illustrated with examples from the great tit, *Parus major. Population Biology and Evolution*. (Eds K. Wohrmann & V. Loeschcke), pp. 67–79. Springer-Verlag, Berlin.

Nur, N. (1984a) The consequences of brood size for breeding blue tits. I. Adult survival, weight change and the cost of reproduction. *Journal of Animal Ecology*, **53**, 479–496.

Nur, N. (1984b) The consequences of brood size for breeding blue tits. II. Nestling weight, offspring survival and optimal brood size. *Journal of Animal Ecology*, **53**, 497–517.

Owen, M. (1980) *Wild Geese of the World*. Batsford, London.

Parker, G.A. (1985) Models of parent-offspring conflict. V. *Animal Behaviour*, **33**, 519–533.

Parry, G. (1982) The evolution of life histories of four intertidal limpets. *Ecological Monographs*, **52**, 65–91.

Partridge, L. & Farquhar, M. (1981) Sexual activity reduces lifespan of male fruitflies. *Nature*, London, **294**, 580–581.

Perrins, C.M. (1964) Survival of young swifts in relation to brood-size. *Nature*, London, **201**, 1147–1148.

Peters, R.H. (1983) *The Ecological Implications of Body Size*. Cambridge University Press, Cambridge.

Phillipson, J. (1966) *Ecological Energetics*. Edward Arnold, London.

Pianka, E.R. (1970) On *r*- and *K*- selection. *American Naturalist*, **104**, 592–597.

Prader, A., Tanner, J.M. & von Harnack, G.A. (1963) Catch-up growth following illness or starvation. *Journal of Paediatrics*, **62**, 646–659.

Pressley, P.H. (1981) Parental effort and the evolution of nest-guarding tactics in the three spine stickleback, *Gasterosteus aculeatus. Evolution*, **35**, 281–293.

Regal, P.J. (1977) Evolutionary loss of useless features: is it molecular noise suppression? *American Naturalist*, **111**, 123–133.

Reiss, M.J. (1985) The allometry of reproduction: why larger species invest relatively less in their offspring. *Journal of Theoretical Biology*, **113**, 529–544.

Reynoldson, T.B. (1983) The population biology of Turbellaria with special reference to the triclads of the British Isles. *Advances in Ecological Research*, **13**, 235–326.

Reznick, D. (1985) Costs of reproduction: an evaluation of the empirical evidence. *Oikos*, **44**, 257–267.

Ricklefs, R.E. (1969) Preliminary models for growth rates in altricial birds. *Ecology*, **50**, 1031–1039.

Ricklefs, R.E. (1982) A comment on the optimization of body size in *Drosophila* according to Roff's life history model. *American Naturalist*, **120**, 686–688.

Ricklefs, R.E. (1984) The optimization of growth rate in altricial birds. *Ecology*, **65**, 1602–1616.

Robertson, R.J. & Bierman, G.C. (1979) Parental investment strategies determined by expected benefits. *Zeitschrift fur Tierpsychologie*, **50**, 124–128.

Roff, D. (1981) On being the right size. *American Naturalist*, **118**, 405–422.

Rose, M.R. (1983) Theories of life-history evolution. *American Zoologist*, **23**, 15–23.

Rose, M.R. (1984) The evolutionary route to Methuselah. *New Scientist*, **103**, 15–18.

Ruse, M. (1979) *The Darwinian Revolution*. University of Chicago Press, Chicago and London.

Sacher, G.A. & Staffeldt, E.F. (1974) Relation of gestation time to brain weight for placental mammals, implications for the theory of vertebrate growth. *American Naturalist*, **108**, 593–615.

Sastry, A.N. (1983) Ecological aspects of reproduction. *The Biology of Crustacea Vol. 8* (Eds F.J. Vernberg & W.B. Vernberg), pp 179–270. Academic Press, New York.

Schaffer, W.M. (1974) Selection for optimal life histories: the effects of age structure. *Ecology*, **55**, 291–303.

Schmidt-Nielsen, K. (1984) *Scaling: Why is animal size so important?* Cambridge University Press, Cambridge.

Schoenheimer, R. (1946) *The Dynamic Steady State of Body Constituents*. Harvard University Press, Cambridge, Mass.

Scriber, J.M. & Feeny, P. (1979) Growth of herbivorous caterpillars in relation to feeding specialization and to the growth form of their food plants. *Ecology*, **60**, 829–850.

Sibly, R.M. (1981) Strategies of digestion and defecation. *Physiological Ecology: An evolutionary approach to resource use* (Eds C.R. Townsend & P. Calow), pp 109–139. Blackwell Scientific Publications, Oxford.

Sibly, R. & Calow, P. (1982) Asexual reproduction in Protozoa and invertebrates. *Journal of Theoretical Biology*, **96**, 401–424.

Sibly, R. & Calow, P. (1983) An integrated approach to life-cycle evolution using selective landscapes. *Journal of Theoretical Biology*, **102**, 527–547.

Sibly, R. & Calow, P. (1984) Direct and absorption costing in the evolution of life cycles. *Journal of Theoretical Biology*, **111**, 463–473.

Sibly, R. & Calow, P. (1985) Classification of habitats by selection pressures; a synthesis of life-cycle and r/K theory. *Behavioural Ecology* (Eds R.M. Sibly & R.H. Smith) pp 75–90. Blackwell Scientific Publications, Oxford.

Sibly, R.M. & Calow, P. (1986) Growth and resource allocation. *Evolutionary Physiological Ecology* (Ed. P. Calow). Cambridge University Press, Cambridge, UK.

Sibly, R., Calow, P. & Nichols, N. (1985) Are patterns of growth adaptive? *Journal of Theoretical Biology*, **112**, 553–574.

Sibly, R. & McFarland, D. (1976) On the fitness of behaviour sequences. *American Naturalist*, **110**, 601–617.

Smith, R.J. (1980) Rethinking allometry. *Journal of Theoretical Biology*, **87**, 87–111.

Sokal, R.R. (1970) Senescence and genetic load; evidence from *Tribolium*. *Science*, **167**, 1733–1734.

Southwood, T.R.E. (1977) Habitat, the templet for ecological strategies. *Journal of Animal Ecology*, **46**, 337–365.

Southwood, T.R.E. (1978) *Ecological Methods: With particular reference to the study of Insect Populations*. (2nd Edition). Chapman & Hall, London.

Stearns, S.C. (1977) The evolution of life history traits: a critique of the theory and a review of the data. *Annual Review of Ecology and Systematics*, **8**, 145–171.

Stearns, S.C. (1980) A new view of life-history evolution. *Oikos*, **35**, 266–281.

Stearns, S.C. (1983) The influence of size and phylogeny on patterns of covariation among life-history traits in the mammals. *Oikos*, **41**, 173–187.

Stearns, S.C. (1984) The effects of phylogeny on patterns of covariation among life-history traits of lizards and snakes. *American Naturalist*, **123**, 56–72.

Stebbing, A.R.D. (1981) The kinetics of growth in a colonial hydroid. *Journal of the Marine Biological Association*, U.K., **61**, 35–63.

Sutherland, W.J., Grafen, A. & Harvey, P.H. (1986) Life history correlations and demography. *Nature*, London, **320**, 88.

Taghon, G.L. (1982) Optimal foraging by deposit-feeding invertebrates: roles of particle size and organic coating. *Oecologia*, (Berlin), **52**, 295–304.

Taghon, G.L., Self, R.F.L. & Jumars, P.A. (1978) Predicting particle selection by deposit feeders: a model and its implications. *Limnology and Oceanography*, **23**, 752–759.

Taylor, C.R., Maloiy, G.M.O., Weibel, E.R., Langmon, V.A., Kamau, V.M.E., Seeherman, H.J. & Heglund, N.C. (1981) Design of the mammalian respiratory system. III. Scaling maximum aerobic capacity to body mass: Wild and domestic mammals. *Respiration Physiology*, **44**, 25–37.

Taylor, H.M., Gourley, R.S., Lawrence, C.E. & Kaplan, R.S. (1974) Natural selection of life history attributes: an analytical approach. *Theoretical Population Biology*, **5**, 104–122.

Tenney, S.M. & Remmers, J.E. (1963) Comparative quantitative morphology of the mammalian lung: diffusing area. *Nature*, London, **197**, 54–56.

Tinkle, D.F. (1969) The concept of reproductive effort and its relation to the evolution of life histories in lizards. *American Naturalist*, **103**, 427–434.

Trivers, R.L. (1974) Parent-offspring conflict. *American Zoologist*, **14**, 249–264.

Tuomi, J., Hakala, T. & Haukioja, E. (1983) Alternative concepts of reproductive effort, costs of reproduction and selection in life-history evolution. *American Zoologist*, **23**, 25–34.

Vance, R.R. (1973a) On reproductive strategies in marine benthic invertebrates. *American Naturalist*, **107**, 339–352.

Vance, R.R. (1973b) More on reproductive strategies in marine benthic invertebrates. *American Naturalist*, **107**, 353–361.

Ward, S.A., Wellings, P.W. & Dixon, A.F.G. (1983) The effect of reproductive investment on pre-reproductive mortality in aphids. *Journal of Animal Ecology*, **52**, 305–314.

Wardle, R.A. & McLeod, J.A. (1952) *The Zoology of Tapeworms*. University of Minnesota Press, Minneapolis.

Ware, D.M. (1975) Growth, metabolism and optimal swimming speed of pelagic fish. *Journal of Fisheries Board of Canada*, **32**, 33–41.

Watt, W.B. (in press) Power and efficiency as indices of fitness in metabolic organization. *American Naturalist*.

Weiss, P. & Kavanau, J.L. (1957) A model of growth control in mathematical terms. *Journal of General Physiology*, **41**, 1–47.

Westerterp, K. (1977) How rats economize — energy loss in starvation. *Physiological Zoology*, **50**, 331–362.

Whittaker, R.H. (1975) The design and stability of plant communities. *Unifying Concepts in Ecology*. (Eds W.H. Van Dobben & R.H. Lowe-McConnell) pp 169–181. D.W. Junk, The Hague.

Wieser, W. (1985) A new look at energy conversion in ectothermic and endothermic animals. *Oecologia*, (Berlin), **66**, 506–510.

Wilbur, H.W., Tinkle, D.W. & Collins, J.P. (1974) Environmental certainty, trophic level and resource availability in life history evolution. *American Naturalist*, **108**, 805–817.

Williams, G.C. (1966) *Adaptation and Natural Selection*. Princeton University Press, Princeton.

Williams, G.C. (1975) *Sex and Evolution*. Princeton University Press, Princeton, N.Y.

Winberg, G.C. (1956) Rate of metabolism and food requirements of fishes. *Fisheries Research Board of Canada Translation Series* No. **194**, 1–253.

Woollhead, A.S. (1983) Energy partitioning in semelparous and iteroparous triclads. *Journal of Animal Ecology*, **52**, 603–620.

Woollhead, A.S. & Calow, P. (1979) Energy partitioning strategies during egg production in semelparous and iteroparous triclads. *Journal of Animal Ecology*, **48**, 491–499.

Wootton, R.J. (1984) *A Functional Biology of Sticklebacks*. Croom Helm, London.

Wootton, R.J. (1985) Energetics of Reproduction. *Fish Energetics: New Perspectives*. (Eds P. Tytler & P. Calow) pp 213–254. Croom Helm, London & Sydney.

Wright, S. (1932) The roles of mutation, inbreeding, crossbreeding and selection in evolution. *Proceedings VIth International Congress of Genetics*, **1**, 356–366.

Author index

Subject index

Note: Page references in italic refer to tables and figures, those in bold refer to mathematical examples.

DATE DUE

GAYLORD			PRINTED IN U.S.A